Team Spirit

A Field Guide to Roots Culture

Geoff Pevere

Doubleday Canada Limited

Canadian Cataloguing in Publication Data

Pevere, Geoff
Team spirit

ISBN 0-385-25808-9

1. Roots (Firm). 2. Clothing trade – Canada.
3. Fashion merchandising - Canada. I. Title.

HD9940.C254R66 1998 338.7'687'0971 C98-931190-2

Care has been taken to trace ownership of copyright material contained in this book. The publisher will gladly receive any information that will enable it to rectify any reference or credit line in subsequent editions.

Project Manager: Gaynor Fitzpatrick
Editor: Charis Wahl
Text and Cover Design: Counterpunch/Linda Gustafson
Cover stitching: Roots
Photo Research: Robyn Craig

Published by Doubleday Canada Limited
105 Bond Street
Toronto, Ontario
M5B 1Y3

FRI 10 9 8 7 6 5 4 3 2 1

This book is dedicated to the memory
of Grace, Jack, Iva and Roy. My roots.

Acknowledgments

Not just about team spirit, this book is also a product of it. Such is the nature of the publishing game that I just get to take most of the credit.

For their support, patience and unfailing good will, I'd like to thank Dara Rowland and John Pearce of Doubleday Canada, and project manager Gaynor Fitzpatrick. Thanks also go to book designer, Linda Gustafson.

The manuscript was read at various stages by Stephen Cole, Claire Davey, Greig Dymond and Kevin McMahon, all of whom provided invaluable and incisive feedback. Moreover, they spoke up when something sucked.

My editor, Charis Wahl, deserves special mention, as she seemed incapable of making suggestions that didn't render the bad and ugly into something good. And boy, does she know her camp culture.

Special thanks are due to my agent (and avid reader) Dean Cooke, who always gives one hundred per cent, but only takes fifteen.

At Roots Canada, my thanks go out to Michael Budman, Don Green, Stephanie Johnston and Raymond Perkins. They gave me full access to whatever I needed, and never asked what the hell for. They even returned most of my calls. For a company as control-crazy as this one, letting me run loose cannot have been easy.

Finally, special thanks to my family, Emma and Claire Davey, for enduring far too many dinner-table conversations about the negative-heel shoe. The weirdo loves you.

Contents

**Coming out in style:
the Canadian Olympic
team makes its debut,
Nagano, 1998.**

Introduction
A Cap for Canada

"We were the first people to put 'Canada' on clothing."
> — *Don Green, co-founder and co-owner, Roots Canada*

"We've been able to give Canada an identity that we're proud of."
> — *Michael Budman, co-founder and co-owner, Roots Canada*

"Are those Roots guys smart or what?"
> — *Susan G. Cole,* Now

To see something clearly, sometimes you have to take it away from home.

Freed of old associations and alliances, unburdened by local prejudice and held up to unfamiliar light, things take on a sharpness they just don't have on home turf. It works for old clothes and lovers, but it holds for countries, too. Such was the view of Canada offered by the 1998 Olympic Winter

Bound for glory: thousands in Ottawa to board buses for the unity rally in Montreal. October 27, 1995.

Games in Nagano, Japan. By the Games' end, the home of the beaver was looking dramatically different. And all because of a hat.

The final Winter Olympics of the millennium, the Nagano Games provided one of the most revealing glimpses of the state of the Canadian nation since the squeaker referendum on Quebec's separation in 1995. On that occasion, thousands of mostly anglo-Canadians boarded Montreal-bound buses to appear in one of the most stirring photo opportunities in the country's recent history. Flooding the streets of the city, watched by the unblinking eye of national television, the anti-separatists waved countless tiny red-and-white Canadian flags in the very heart of the city where mailboxes once exploded and bricks had been hurled at Prime Minister Pierre Trudeau, where the army had once commandeered bloodstained streets and General de Gaulle had vived "le Québec libre." If Canadian disunity had a heart, surely it was in Montreal.

Mutual attraction: Roots has produced Olympics-related products since 1976; but it wasn't until 1997 that they became the official Olympic outfitter.

Toronto '96

And yet, when the buses arrived, the cameras rolled and the flags were waved, the sum total was a visual image of almost stirring national togetherness, and never more so than in the widely reproduced image, taken from above, of the crowd holding an impossibly immense flag over their heads — all those fresh-faced, multi-cultured youths dressed in pastel winterwear, waving the maple leaf as if the collective wave of some tourist-shop magic wand were going to hold this troubled confederation together.

But hold on a sec. That was cynical, wasn't it? Cynical and, let's face it, beside the point. For what all those young Canadians served to do, with all their waving, singing and cheerleading for a unified country, was to provide an image that rendered reality irrelevant. As far as anyone watching the TV coverage was concerned, the country was bonding by the sheer adhesive power of the positive televisual image. It was a big, warm hug for an ailing nation, a regenerative embrace consisting of bright colours, big smiles and bouncy songs. So what if, as a country anyway, we were still screwed up? At least we looked happy, and by this point in human history, is there any better way of being something than by looking so? Is there any difference?

One could have asked the same questions while watching the opening ceremonies of the 1998 Winter Olympics. For, like the Montreal referendum rallies for national unity, the opening of the Olympics proved a terrific, if not altogether accurate, photo op for a unified Canada: the first appearance of the Canadian Olympic team, outfitted in the

Goes well with gold:
the Canadian Curling Team
at Nagano, 1998.

Goes well with green, too:
Ross Rebagliati wins big.

alpine-skiing-meets-snowboarding uniform designed for them by the Canadian-based clothing company called Roots.

Once again, the predominant colours were red and white. The colours of candy canes, Santa Claus, the Coca-Cola label, the Detroit hockey uniform and the Canadian flag, red-and-white is a feel good visual vibe; and the Roots Olympic uniforms capitalized on it: funky but formal, and topped off by the insouciantly backwards-worn "poorboy" cap, which would instantly become a fashion phenomenon, the uniforms – described by the *The Globe and Mail* as "a heady combination of hype and authenticity" – proved as irresistible to the world as they were to the country they represented. And, for purposes of brand recognition, it didn't hurt that "Roots" was nearly as visible on the clothing as "Canada" was, particularly on that hat. "We were given the exact specifications of the size of the Canadian Olympic Association logo, the maple leaf with the five rings," Roots co-founder Michael Budman explained to a reporter, "as

Hope on ice: Elvis Stojko
Roots for the home team.

well as the size the Roots logo could appear on the clothing."

Within hours of that appearance, the uniforms were attracting offers from all corners of the Olympic village: people wanted to know where they could get those spiffy-looking Canadian duds and, according to some reports, were even willing to buy them off the backs of Canadian athletes.

Meanwhile, back home, the managers and staff of the nearly 100 Roots outlets across the country braced themselves for a full-fledged retail rush. In a characteristic stroke of marketing savvy, Roots had made certain that most of this irresistible-looking stuff was available in Roots stores, with a twelve percent royalty going to the Canadian Olympic Association. Consumer demand for all Olympic-issue Roots clothing was both instantaneous and overwhelming, but nothing compared to the race for the cap. "It's been insane," the manager of the Roots store in Toronto's Eaton Centre told a newspaper reporter shortly after the opening ceremonies. "Every time we get them in, they sell out in fifteen minutes." It was the same across the country: as fast as the hats rolled off the production lines and into the stores, they flew out the door.

It reveals something about the power of the televisual image that the cap didn't take off until the opening ceremonies were televised in February 1998, even though it had been sitting in Roots stores since the previous November. For when it appeared on TV, it became something much bigger and more powerful than a piece of eye-grabbing headgear on a store shelf. It became a symbol: a symbol of a

**Poorboy fever:
the cap that turned all heads.
Red, that is.**
from top **Kelly Lynch, Mark
Messier, Robbie Robertson,
and Pamela Anderson.**

from top **Mike Myers;
Ross Rebagliati and the
Green family; Russell deCarle
of Prairie Oyster; and Ross,
again, with the Budman family.**

country united in collective endeavour; a symbol of a country that was youthful, happy, healthy and hip; a symbol of a country that, to all appearances anyway, was about to kick major international ass on the rinks and slopes of Nagano.

Never mind that the symbol might have contradicted reality – particularly on the unity issue. That's what symbols are for: they reorganize reality into something altogether more attractive and appealing. The symbolic value of the Olympic team's cap and uniform was clear to Roots' owners. "The Canadian team looked great at the opening ceremonies," Budman said in an early post-Olympic interview. "Much better than the Americans. They were unified."

Unified. Better than the Americans. Geez, you couldn't have come up with a more seductive line for Canadians in 1998 – Roots' twenty-fifth year in business – than if you'd got the whole country drunk and promised it a honeymoon in Florida. If Roots had dressed the anti-referendum forces in Montreal, the "No" vote might have been less a squeaker than a slaughter.

(Besides, as certain members of Parliament have learned since, a flag waved without flair is just more wind. Undoubtedly influenced by the referendum rally, Canadian Heritage Minister Sheila Copps attempted in 1996 to distribute a million Canadian flags to citizens across the country, most of whom were audibly unimpressed that the scheme cost 23 million taxpayers' dollars. And in spring '98, doubtlessly influenced by Nagano, the official Opposition in the House of Commons, the Reform Party, attempted to

make a federal case out of placing dinky little Canadian
flags on Members' desks. Once again, the electorate was
unmoved, and the Reformers were ridiculed for having
nothing better to do than make a fuss over toy flags. Maybe
if the Reformers had insisted everybody in Parliament wear
those sharp-lookin' Roots poorboys...)

Casual-chic, the Roots team uniform was funky and
functional, topped by what would become one of the indelible visual tropes of the Games themselves: that poorboy
cap, inspired by hiphop but customized for Team Canada,
seemed to make daily front-page appearances around the
world. And when it did, whether it was adorning the scalp
of figure-skating champion Elvis Stojko, hockey super-
duperstar Wayne Gretzky, or off-again-on-again gold-win-
ning snowboarder Ross Rebagliati, it was impossible to miss
two words embroidered on the brim: "Roots" and "Canada."
Clearly, somebody at Roots had been thinking about the
televisual potency of Canada's national colours, a lesson the

The Canuck as news.

referendum rallies brought home with the force of an Eric Lindros crosscheck.

But there were other, more substantial if subterranean similarities between the referendum rallies and the Roots-issue Olympics. For like the happy faces of those hale young flag-wavers, the Roots cap cheerfully hid a host of domestic scabs and scratches: seeing it for the first time — or even the second, seventh or thirty-fifth — it was all too easy to forget that, in 1998, Canada was every bit as fractious as ever, what with a plummeting dollar, federal-provincial mudslinging, radical institutional downsizing and (those maple leafs notwithstanding) the chronically vexed issue of Quebec separation. (Indeed, the night before the opening ceremony, as the Canadian Olympic Team gathered to watch a state-sponsored pep-rally video, Canada's most persistent national sore spot threatened to infect the kick-off party itself: barely any French was spoken on the official tape, triggering a barrage of accusations that the Canadian Olympic Association was insensitive to French-speaking Canadians. Oops.)

But all this merely provided the uniform with an Olympic-calibre challenge, one it would meet with golden results. For, like all successful uniforms, and like all successful mythmaking in the television era, the uniform — particularly the cap — provided an image of Canadian unity that made all the squabbling back home recede like fading radio static. For two weeks at least, the Roots cap and uniform conveyed a Canada that was together in all senses of the word, and cool in two. "We used a jacket style that combined

vintage sportswear and contemporary snowboarding," Budman told the Vancouver *Province*. "We wanted everyone to look like a unified team, whether they were curlers, skaters, skiers or snowboarders."

For many people, that cap was the coolest. Indeed, the Olympic poorboy quickly demonstrated signs of being in the late-Nineties what the negative-heel craze had been to Roots in the Seventies and the Roots Beaver Athletics sweatshirt in the Eighties: a bona fide consumer bonanza, the kind of craze that means you've just gotten that much bigger and better and more secure. The opposite, some might say, of the country itself.

Still, if there were an irony to the retailing victory of the Olympic cap, it was one that could not have been lost on the hopelessly jockstruck imaginations of Michael Budman, 52, and Don Green, 48, co-founders and owners of Roots Canada, the most successful clothing designer in the history of a country not famous for clothing design, and a corporate

entity whose success is founded in no small part on the cagey marketing of a mythical "Canada" to a consumer constituency that, even before Nagano, had spread well beyond the woodsy hinterlands of the Great White North. These two, as we will soon discover, played key roles in the popularization of "team" clothing as global urban streetwear. American expatriates, whose profound love of their adopted country (not to mention their fashion sense) is

**The grateful one:
Wayne Gretzky leaves Nagano,
taking the jacket with him.**

inextricable from the culture of sport, Budman and Green were as charged up with Olympic hopefulness for certain Canadian athletes as were any of their born-and-bred fellow Canucks — particularly the Canadian athletes who plied their sport on skates. In the weeks leading up to the games, Canadian billboards were plastered with the Roots-wrapped image of gold-hopeful Elvis Stojko, and Canadian hockey team hero Wayne Gretzky was already making public appearances in Roots athletic leisurewear. Needless to say, Green and Budman, "the Roots Boys," as they are invariably called, seemed to have a whole lot invested (emotionally and otherwise) in the gold-calibre success of Canada's most prized and prominent hopefuls.

Besides, there was something everybody knew but no one dared talk about during the high-tension atmosphere of the Games themselves: that there was more at stake for our superstar skaters than for just about the rest of our Olympic efforts combined. Not because what the skating stars did was more important or valuable, and certainly not because Stojko and Gretzky were better athletes than anyone else, but because, let's face it, one of Canada's most essential points of national pride is the idea that, no matter what everybody else might have on us, we skate like nobody's business. Take away our skates and, in a manner of speaking, you take away our balls. You remove our pride and our sense of who we are. Needless to say, this becomes critical only when the country seems to be squabbling itself into splinters — that's when you need the comforting safety net of a national mythology, and

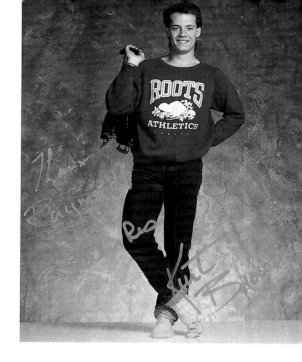

Good things come to those who skate: Kurt Browning sans blades avec Roots.

Excellent free stuff, man: Mike Myers with Budmans and Greens.

21

Putting a lid on it.
above Céline Dion;
below k.d. lang, a price
still on her head.

that's why Canada landed hard when that mythology no longer proved strong enough to keep our national dreams from ripping through the strands.

Which is why the triumph of the cap is all the more remarkable, not just for Roots but for our understanding of the state of Canadian nationalism at the brink of the next century. For the fact was, while the worst *did* come true, while the failure to bring home the gold by both Gretzky and Stojko (who came down with an irony-enriched groin injury) fed our nightmares of national inadequacy, the mythical value of those damned caps skyrocketed. Without them, we'd probably have limped away from Nagano a much more gravely wounded nation, for they (along with the record-breaking fourteen *other* medals Canadians won) allowed us to walk proudly with the bright symbol of something uniquely, distinctly and *proudly* Canadian right up there on our heads, for the whole world to see. Heck, it even says so: "Canada," right next to "Roots."

And not just to see but to *buy*, for this is not a book about a country or a clothing company alone, but about the symbiotic fusion of the two, so that fashion and nationalism, Canada and Roots, culture and retailing are as hopelessly interdependent as, well, Michael Budman and Don Green, even after all these years.

On one level then, it's about how Roots — which has successfully packaged an idea of Canada during a particularly troubled stretch of the country's history and is now successfully selling that idea around the world — represents a fascinating shift in the way the country thinks about itself as a country. At the century's end, the made-in-Canada success of Roots suggests that the most productive context for the marketing and selling of national pride is no longer the public sphere of education or government, but the private playground of the mall, the place of fantasy and possibility, where just about everything can be had for a price, even, it would seem, a sense of belonging.

A few words about what this book is not. It isn't a business profile, for the numbers tell only a small part of the Roots story. It isn't a biographical portrait of Don Green and Michael Budman, although, as you will see, their story is inextricable from that of the company they created and continue to oversee with an almost parental intensity. It isn't an official or authorized Roots product, although it could not have been written without the cooperation of the company, which generously opened both its doors and files to me for the purposes of research, and which never tried to influence what was being written about it. It is, instead, an attempt to study Roots in a cultural context, which is the arena where, to float an analogy Budman and Green might use, the company has really played the best game and scored the most points.

Canajan, eh? One symbol serves another: unlike the 'Mountie,' below, Annie Perrault is the real thing.

23

Still eager after all these years:
Heather Cooper's original
beavers away for Roots.

Feet First

The Negative Heel Affair

"Be kind to feet. They outnumber people two to one."
— *negative-heel ad tag-line*

Like the story of Canada, the Roots story has any number of beginnings, but this is one of the best. It opens in a monastery in Santos, Brazil, circa 1957.

A Danish-born clothing designer and yoga teacher recovering from an operation, Anna Kalsø was drawn to this remote South American monastery to study yogic breathing techniques. Struck by the impeccable posture of barefoot Brazilian natives — and recently converted to the spinal benefits of the lotus position — Kalsø began contemplating the circuits of wellness connecting the soul, the posture, the feet, and the ground. For most people, these are short-circuited — by carpeting, concrete, inertia, stiletto heels or maybe Western civilization in general. For Kalsø, drawing fresh air deep into her rejuvenating body and soul, one thing about the circuitry of good health became dead clear: it started where the ground meets the body, and where each

25

How Roots give your feet a good feeling, then send it up your spine.

To see the idea behind Roots, take a side-view look at the shoe. Instead of a heel to lift you up and tilt you forward, you'll find a one-piece base to plant you firmly in touch with Mother Earth. Roots, you see, work very much like roots. And if you take a side-view look at the human foot, you'll see why they work as well as they do. Your heel is the lowest part of your foot, so in Roots it sits in the lowest part of your shoe.

Suddenly you stand straighter as additional muscles in the back of your legs and the small of your back spring to life to help hold you up and move you around. Now consider that recess in your sole called the arch. If you spend a good deal of time on your feet, unsupported arches can sag and may fall out of shape altogether. (This is why in those pre-cruiser days a police-man was known as a flatfoot.) To help prevent your arches from falling, Roots are contoured to support them. There's a smaller recess between the balls of your feet which Roots will take care of as well.

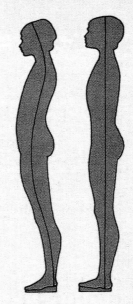

Near the front, you'll notice the sole is curved like a rocker. In normal walking, your weight lands first on your heel, shifts along the outer side of your foot, then diagonally across to your big toe which springs you off on your next step. The rocker idea simply makes that transfer of weight a little easier, which makes each step a little less tiring.

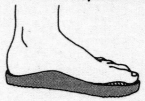

All told, Roots bring a good, natural feeling to man's somewhat un-natural custom of treading hard floors and city sidewalks. Roots are designed and made in Canada; and at the heart of our production are two generations of cobblers (a father and three sons) who cling to the premise that good quality footwear must still be made largely by hand. The way we feel about making Roots has a lot to do with the way you'll feel wearing them.

roots
NATURAL FOOTWEAR

City feet need Roots.

step reminds us (or *should*) of our primeval connection to the earth.

The problem was, you weren't going to convince too many Florsheimed first-world types — however spiritually desperate or spinally challenged — of the benefits of walking barefoot, particularly on ice and asphalt. So, if you couldn't change the ground, how about the way you walked on it? And that's how Anna Kalsø got the idea for "The Earth Shoe." Or so the story goes.

Proceeding on the podiatrically unorthodox principle that the natural and most healthful position for the walking foot placed the heel lower than the toe, Kalsø set out to develop a shoe that would replicate the sensation and benefits of walking barefoot. Some several hundred miles and ten years later, she had personally road-tested her idea to market-readiness.

Patenting the design and Earth Shoe name, Kalsø sold the vaguely ducklike shoe exclusively from a small shop in Copenhagen. It being the late Sixties, word of the enterprise got around; and one day a movie producer named Ray Jacobs wandered into Kalsø's shop. Seems that Ray's wife Ellie had been suffering from persistent lower back pain, so they decided to check out this shoe they'd been hearing so much about.

After a few days of walking around in Earth Shoes, Ellie and Ray were thrilled — Ellie by the apparent relief from spinal discomfort, Ray by a potential countercultural marketing bonanza. So they asked Anna about taking the recessed-heel idea to the States. Eventually Anna agreed, but only after subjecting the Jacobses to a battery of spiritual

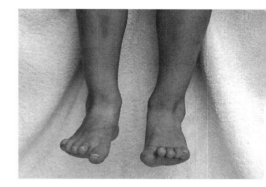

Birth of a notion:
"Your feet are your roots."

Don Green goes native in Negril, Jamaica, 1971: is the negative-heel inspiration looming on the horizon?

fitness tests — personal interviews, astrological charts — and making sure that she kept at least one foot (so to speak) in the coming Earth Shoe Bonanza.

The first Jacobs Earth Shoe outlet opened in 1970 in Manhattan, its unveiling timed to coincide with "Earth Day." A smash success by both establishment and anti-establishment standards, the modest orthopedic empire soon spread to more than forty outlets nationwide, each staffed by people who, according to *The Berkeley Monthly*, "share a deep common interest in health, yoga, natural foods, bio-energetics and 'making the most of life.'" Through the countercultural press particularly, the physical and spiritual benefits of the negative-heel shoe were extolled by podiatrists and alternative-lifestyle gurus alike; and soon the indisputably clunky-looking footwear was adorning tuned-in, turned-on and dropped-out feet across the nation. Not surprisingly, two of those feet belonged to Don Green.

He was an upper middle-class drop-out son of an auto parts manufacturer, whose interest in things healthy was matched only by his interests in hockey and making a living on his own terms. Green and Michael Budman — both ex-Detroiters with draft deferments and therefore in the unusual position, for draft-age American males, of being in Canada by choice — were looking to start a business in Toronto that would allow them to continue the laid-back life to which they'd become accustomed. The two had met as teenagers in 1962, and over the years had forged a bond as tightly stitched as the leatherwear they'd soon be famous for.

Budman, like Green, had fallen hard for Canada while a camper at Algonquin Park's Camp Tamakwa; he had moved to Toronto when the streets of Detroit became a riot zone in the late Sixties, and his stint as a high school teacher had devolved into a gig hauling fruit crates through the Ontario Food Terminal. Green had recently returned from bumming in Jamaica after dropping out of Michigan State.

Clearly, they needed something to do — provided it wasn't too strenuous. As Green revealed to a TV interviewer years later: "We said to ourselves, 'What is something we could do where we could get a lot of time off?' "

Having studied the retail potential of everything from yogurt and waterbeds to flowers and fruit, Green suggested they get into the negative-heel trade, which was already showing signs of the boom to come. So they went to New York to see Ray Jacobs about acquiring the Canadian franchise rights to the Earth Shoe.

At first it looked good for the ex-Tamakwans, but things fell through. "He was a corporation type," Green told *People* of Jacobs. "The vibes weren't right." Undeterred, and probably bereft of alternatives, Budman and Green decided to go it alone. Why not design their own version of the Earth Shoe and sell it themselves? To this end, Green got to work designing a more user-friendly, less hippie-utilitarian variation on Kalsø's original: a less radical incline between toe and heel, a soft, all-natural interior, and a design with crossover potential beyond the yoga-yogurt set. Not to put too fine a point on it, a dorked-down Earth Shoe.

Bowing to tradition, 1966: former Tamakwa canoe instructor Michael Budman would soon be floating a concept.

Sport Root · Mary Jane · Sun Root · T-Strap Root · Roots Clog · Main Root · Moccasin Root · Pump Root · City Root · Spo

The Original Root
The Barefoot & The Wedge

Kind of ugly, but as comfortable as bare feet on a beach.

Freeze that frame. We've reached a key moment in Roots history, the first occasion when the company would profit enormously from the careful modification of something already there. Roots is not a fashion innovator, but an expert marketer of the traditional and familiar to the conservative mass market. Thus, it didn't invent the negative heel, but modified it for wider appeal than the hippie boutique market, and sold it with a zeal and intelligence that *was* new. In the coming decades, it would happen again with the Roots Beaver Athletic sweatshirt and the poorboy cap, both of which had been around for decades by the time Roots turned them into full-fledged fashion phenomena. By catching a wave at precisely the right time and knowing just how to ride it, Roots has consistently surfed in the black.

Applying the basic design to several shoe styles – desert boot, moccasin, city and sport – Green was ready to put the product into production. All he needed was a factory that actually made shoes. Throwing a $15,000 loan from Don's father – like Budman's dad, a prosperous Detroit entrepreneur – into their pooled Bar Mitzvah savings, the boys went to the first shoe manufacturer listed in the Toronto Yellow Pages. But the shoe giant Bata didn't bite – imagine the pitch meeting between the buttoned-down Toronto footwear executives and these callow hippies with wild hair and weird shoes.

Letting their fingers do the walking, Budman and Green dialled the next shoemaker listed in the book. Boa Shoe Company was a tiny family operation run by a Polish expatriate whose family, in the footwear business for eight generations, had once made boots for Czar Nicholas ii. Now Jan Kowalewski, surrounded by his shoemaking sons Henry, Richard, Stanley and Karl, was listening to two young strangers with a dog in tow and a notion to make shoes that would make people feel like they were standing barefoot in sand. Absurdity notwithstanding, Jan Kowalewski agreed –

Gold in them thar heels: the boys prepare for the short march.

The lease for the first store at 1052 Yonge Street in Toronto. Payments were $280 a month. Within a year, a million dollars' worth of shoes would walk out that door.

on condition he could add his own modifications – to produce 120 pairs of these odd and unCzarlike shoes for these two kids from Detroit. Or maybe it was for the dog, by all accounts a charming beast.

Now Budman and Green had everything except a company name and a store to sell from. "Roots" was found by Don, who kept stumbling across the term in a girfriend's textbooks. Don liked its all-natural, organic ring: "Your feet are connected to the ground, like a tree is," he said earnestly in an interview. (Several, actually.) "Really, your feet are your roots."

With a Heather Cooper-designed beaver logo adorning the doorway, the first Roots store opened on Yonge Street just north of Davenport, mere blocks north of the former hippie mecca of Yorkville.

On day one, August 15, 1973, the store moved seven pairs of the "Roots Shoe" at a hefty $35 a pair. Hardly through the roof. The next Saturday thirty pairs walked out the door, which meant the doors could remain open – for another week.

The following Saturday, for reasons only slightly less fathomable than the Seventies themselves, the ethereal forces of fashion faddism converged above the little shoe store on Yonge Street, and they were smiling.

There were line-ups around the block. The shoes sold out, and there were waiting lists for the customers at the end of the line. Said Budman ten years later, "It was like a gold rush."

Wearing the wares:
Budman and Green look
positive in negative heels.

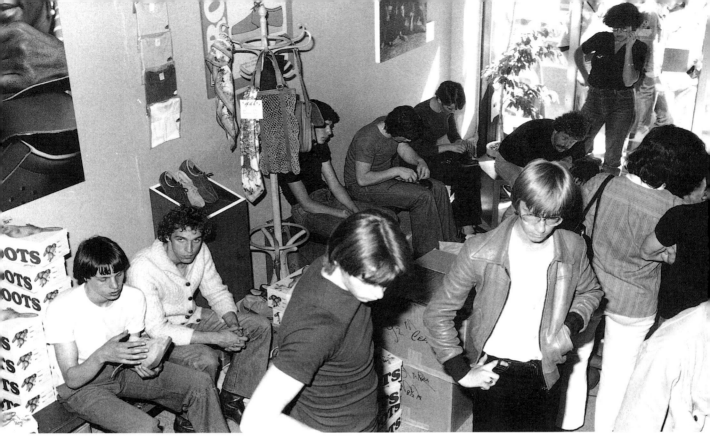

However you choose to characterize it, the Roots shoe — more than the Kalsø original — was a bona fide, out-of-nowhere, standout phenomenon. Indeed, a full accounting of the popular tastes of the Seventies is frankly incomplete without the Roots shoe: it deserves a parking spot in historical posterity every bit as lofty as those reserved for Farrah, Fonzie, Burt and *Dark Side of the Moon*. It was so popular so quickly that, within three years, the Roots-made negative heel was not only dubbed the "the Gucci shoe of the crunchy granola set" by *People* magazine, but it made full-fledged entrepreneurial marketing stars of Michael Budman and Don Green.

Demonstrating the promotional instincts that would quickly set Roots apart from all contemporary competitors,

34

above Have them back by midnight!: Mike and Don play fairy godmother to a model in Amsterdam.

middle Perhaps something a little sportier?: a dirty job but . . .

bottom Swept off her feet by Green and Budman, a Dutch model manages to make everyone smile, including a photo of Jan Kowalewski.

Budman and Green instantly started sending free pairs of the shoes to some of the most famous feet in the world: Paul McCartney, Cher, Elton John, Pierre Trudeau, his wife Margaret — all were photographed wearing their negative heels, and each photo yielded incalculable to-die-for free promotion. Meanwhile, the founders began compensating for the company's lack of advertising resources by acting as their

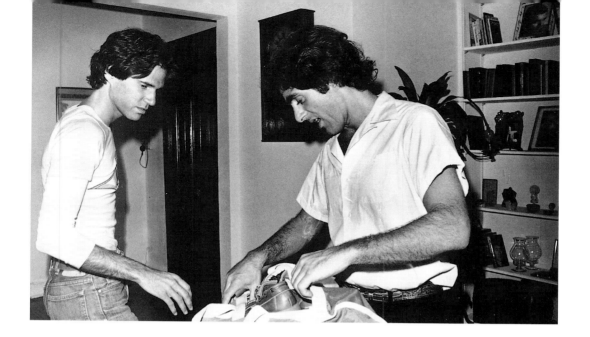

own models. Both tactics worked. By 1975, you'd have had to have been holed up in a Brazilian monastery not to have heard of Roots.

If it were too good to last — and who had time to consider such things with all those hungry feet waiting around the block? — Budman and Green were too busy calculating the next step to fret much. More practical problems loomed: with this volume of shoes being sold, it was obvious that Roots needed its own footwear factory. So the suddenly prosperous owners — whose company was worth more than a million dollars in less than six months — made an offer to Jan and his boys: how would they like to become full-time employees of Roots, applying their eight generations of intercontinential shoemaking expertise to the negative heel alone?

Keeping whatever first impressions he might have had about Budman, Green or their strange little shoe to himself, Kowalewski agreed, and within three months the family was working for Roots exclusively. (They still do.) A new

The Santa Clauses of the North fill a sack with goodies.

plant was commandeered to keep up with demand, and within a year the Kowalewskis were making 2000 pairs of Roots footwear a day.

If the negative-heel boom had been the result of a master plan, there'd still be seminars on the shoe's making and marketing. However, like much of the Roots story — and like the Canadian comedic talent the Roots guys would soon hang out with — it was really blind luck colliding with inspired improvisation. From the far side of success some years on, Green offered this account of Roots' strategy for world domination: "We had absolutely no idea of what we were doing. But we were having so much fun doing it." Or, as Budman told an interviewer on the tenth anniversary of the fortune-making recessed-heel rush, "I have no idea to this day how we ended up making Roots."

Maybe that's because Roots made *them*.

But if the success of the Roots shoe took both consumers and creators by storm, and if Green and Budman are renowned for their faith in the mystical "gut" instinct, the collaboration was hardly calculation-free. For example, while the Kalsø Earth Shoe was promoted on a wave of dubious therapeutic claims, the Roots variation promised only tops in comfort and quality. "Unlike ordinary shoes," reads a 1975 ad, "you stand *in* Roots, not *on* them. There are other casual shoes that look like Roots at first glance. But none of them has the obvious love of good leather and fine boot-making you'll find in a pair of Roots."

The Family Kowalewski, 1974.

Tradition, commitment and craft.

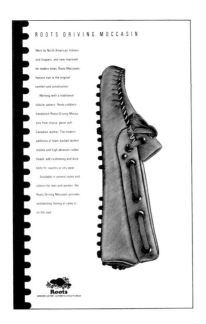

No, not for golfing: the Roots 1982 'driving' moccasin. Beats a portage any day.

The Roots '50's model graces the invitation to Green's 1979 birthday bash. Strangely, the "50's disco party" never caught on.

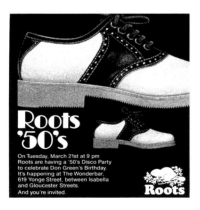

Not once did Roots claim that the shoe was good for you, nor was it recommended for people seeking podiatric relief. You wore them because they felt great, because they were a quality product, and because you were simply too confident and self-actualized to give a toss what other people thought. A good strategic move, for it would be the issue of the Earth Shoe's alleged therapeutic benefits that would bring the negative-heel fad down harder and faster than Evel Knievel over Snake River canyon.

Good thing they'd finally decided to listen to designer Robert Burns, who had challenged an elementary team rule back in 1975 by refusing to don Roots shoes. Asked about it, Burns was simple and direct: "I can't wear the damn things. And besides, you guys aren't in the shoe business. You're in the fashion business. One day soon you're going to have to start making conventional shoes."

Soon was the key. Within a year, Burns' prediction had come terribly true. Ironically, it would be a doctor from the boys' hometown who ripped the sole out of the negative-heel craze. As Budman told an interviewer from the *Detroit Free Press* in 1983, "We had one store in Birmingham that was selling close to 700, 800 pairs a week. All of a sudden some Detroit doctor came out and said the negative heel wasn't good for you. That was it. You could have shot a cannon in the store the next week and not hit anyone."

Good thing then that the two owners had taken Burns' suggestion to heart and had slowly added not just more conventional shoes to the Roots repertoire, but

THE CLASSIC LINE.

This year, Roots has a new line. Same superb styling and craftsmanship above, full rubber sole and heel below.

selected quality leatherwork and clothing as well. Thus, when the Roots shoe craze went downhill, the Roots company was left with something to sell, not to mention a tidy fortune to sell it with.

(A parenthetical pause: the craze died out everywhere but Japan, which remains the only market for which Roots produces negative-heel shoes. It seems a fair trade for electronic pets.)

All they needed was a strategy for life after fad — you know, a game plan. Ironically, for two guys with a passionate love of sport and a Zen-like commitment to the mantra of teamwork, a game plan was the one thing they didn't have. Still, blind luck had put a spring in their step and a fortune in their pockets once before. Surely it would happen again, wouldn't it? Well, wouldn't it?

Roots "Classic":
no sugar, no caffeine, 1979.

TIME
WOUNDS
ALL HEELS

20年前に生まれた"ナチュラル・シューズ"が、ルーツだった。

ROOTS

1970年代初め、アメリカに生まれたナチュラリズムを伝える日本の雑誌は、「ルーツ・カナダ」の"ナチュラル・シューズ"の存在を伝えた。都会人に足元から自然とのかかわりを再認識させた。

A fair trade for electronic pets:
Roots ad for Japan,
the only surviving market
for the negative heel.

It's a wonder of fashion that — Japan notwithstanding — we haven't yet seen a revival of the negative heel. Particularly now, when it would seem, you know, a natural. For not only has this decade witnessed the widespread rehabilitation of what *Spy* magazine called "The Decade That Taste Forgot," but the peculiar footwear fancies of the period — from clunky suede loafers to teetering vinyl platforms — have proven ripe for ironic resurrection.

In the unmemorably named "grunge era," it was a sign of one's alternative-rock credibility that one wore the wide, soft sneakers that many of us shuffled through the Seventies wearing. And by the second half of the Nineties, the impossible had happened: after three decades of reptilian non-evolution, the lowly Wallabee — favoured by biology teachers and audio-visual club treasurers the world over — had attained certified hipness. Don't believe it? Well, there they are: in beige suede splendour, on the cover of one of the decade's biggest albums, the Verve's *Urban Hymns*, adorning the elegantly wasted and unsocked feet of lead singer Richard Ashcroft. I ask you: can that Evel Knievel comeback be far behind?

If the Wallabee — which no Seventies rock star would have been caught dead wearing — can go from hideous to hip, why not the negative-heel shoe? This might have to do with obsolescence: unlike, say the Wallabee or the Converse One Star, which were just shoved to the back of the shelves by more urgent casual footwear fads, the negative heel was pretty well wiped out by late-Seventies medical reports disputing the therapeutic claims made for the recessed heel. From obscurity to triumph to scandal to oblivion: such was Richard Nixon's journey during the Seventies, and so would seem to be the destiny of the negative heel. But even Richard Nixon has enjoyed more positive historical revisionism than the shoe with the backward sole. And he had no soul at all.

Nor is it that the shoe is simply too, um, rooted to its period. For if we've learned anything from Nineties fashion, from the return of the Wallabee to the resurgence of poly-ester, it's that *nothing* is so historically specific that it can't be re-sold wrapped in the quotation marks of irony. The dorkier the better: one of the defining elements of Nineties youth

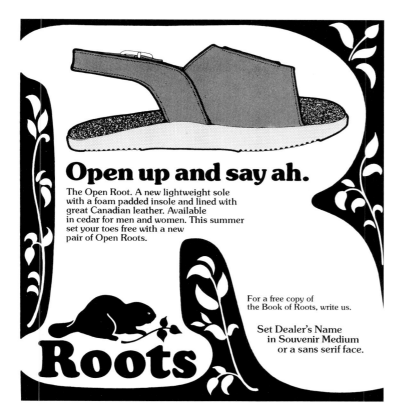

The open Root:
mixing foot and mouth
metaphors is good business.

culture is its deliberate pursuit of precisely those fashions that make oldsters cringe in embarrassment and disgust.

Besides, it's that evocation of a particular moment, that stridently anachronistic nature, that makes clothing items ripe for revision in the first place: that's why they sell poly-cotton *Charlie's Angels* T-shirts, and why it's once again safe to wear collars with Boeing 747 wingspans. If it makes boomers quake, it's cool.

Whatever you want to say about the Roots recessed-heel shoe, its own roots are planted firmly in the moment that is the mid-Seventies. Suspended between a sneaker and a dress shoe, the all-occasion Roots shoe fairly screamed function over fashion. You didn't wear these to look good (otherwise they wouldn't go with everything from a Speedo to a three-piece suit), you wore them to *feel* good; and few periods were more preoccupied with the pursuit of pure, me-first mellow-ness than the Seventies.

And that's exactly how the shoe was sold. It was the choice for people who cared far more about comfort and quality than looking spiffy, for those simply too tethered to the earthly values of health and well-being to perch them-selves on a pair of towering, Elton John-issue platforms. In a word, if the Roots shoe conveyed status, it did so through the time-tested strategy of pitching directly against status. While fashionable platform footwear went ever skyward, the Roots recessed-heel response brought you right down to terra firma. Like the successful anti-fashion campaigns for every-thing from the VW Beetle to Canadian-made Tilley clothes

— which corners the market in garments that are unsightly *and* indestructible — the Roots negative-heel campaign capitalized on anti-fashion fashion. Or, as Huey Lewis observed: "It's hip to be square."

Indeed, in those days being *proudly* square was a measure of self-actualization, a sure sign of the centred self. "They don't look like ordinary shoes," an early Roots ad reads, "which bothers some people. But, if you're secure enough to deal with a few characters who want to know why you're wearing those 'funny-looking shoes' you're going to love the comfort of Roots." There you have it: comfort's conquest of cool, or maybe comfort's moment *as* cool.

Produced by a decade buffeted by cultural cross-currents — health vs. hedonism, fashion vs. function, individual vs. mass, environmentalism vs. consumerism, laid-back vs. in-your-face — the Roots shoe might have flourished only in a time as open, uncertain and oxymoronic as the Seventies, an era that also popularized false-tuxedo T-shirts, fake turtlenecks and, of course, the epochal leisure suit. Even then the shoe flourished only long enough to get a little company called Roots off and running, tilted slightly uphill every step of the way.

Which returns us to our mystery, that conspicuous absence of the recessed-heel shoe from a retro-obsessed fashion market. For if it were the Seventies and only the Seventies that could have created the Earth Shoe craze, then surely there's no better time than now to *recreate* it. The trick is to move quickly, before hip shifts to square all over again.

Denyse Tremblay, not yet married to Don Green, offers up the new negative-heel 'sport' shoe, anticipating the coming cardiovascular craze, 1974.

45

Algonquin:

the promised land.

Heart of Parkness

The Algonquin-Tamakwa Connection

"Here's the truth. Roots is less a company than a summer camp."

> — *Michael Posner*, Toronto Life, *1993*

"Being from Detroit, what stands out in my mind is the whole adventure of getting there. Taking a train to Toronto, a bus and finally a boat to the camp. I remember how beautiful and clean the lake was, and the different clothing that was worn in association with camp."

> — *Michael Budman*

The farther upriver you proceed in the Roots story, the closer you get to Camp Tamakwa, the Algonquin Park kids' camp where Don Green and Michael Budman spent summers as children and teenagers, and where, as campfire legend has it, the Roots concept was born.

Naturally, the best way to get to Tamakwa is by canoe: watching the dawn mist rise above the lake, listening to the

Going my way? You bet you are: Trudeau, buckskin, a canoe and a country. Ottawa, 1983.

distant call of the loon, breathing in tandem with the paddle's pull against clear water. Heaven for the hearty in a made-in-Canada tradition.

The canoe is to Canadian mythology what the wagon train is to the Old West, or maybe the gondola to Venice. A means of transport that not only expresses the geographic personality of the landscape that created it, but a way to the heart of a place's mythic terrain. This is why canoes loom so large in Canada's image of its ideal self. They were an indispensable means of travel for Canada's aboriginal people, and elementary to the establishment of the country's commerce, settlement and industry — they were, in other words, the link between pre- and post-European settlement. Without them, the *coureurs de bois* might have packed it in around Ottawa. Pierre Trudeau, resplendent in buckskin, flannel and chiselled cheekbones, eased a canoe out of the mists of history in the televised version of his *Memoirs* in 1993; and a popular beer campaign — featuring wiseacre *coureurs de bois* — also invoked the mythic power of this perfect machine.

In Canadian schools during the Sixties and Seventies, it was impossible not to be subjected to multiple screenings of a National Film Board of Canada short called *Paddle to the Sea*, about a hand-carved toy canoe that makes its way from one coast to the other, battered and buffeted but unbroken: the rustic essence of Canuck tenacity. Grey Owl, an Englishman who became an international celebrity by impersonating a Canadian Indian during the Depression, made sure for credibility purposes that he was filmed and photographed in

**Not from Detroit:
no assembly required.**

Roots Algonquin

R OOTS ALGONQUIN...clothing and accessories born in the Canadian Wilderness.
Algonquin's majestic hills, sparkling lakes and forest primeval inspired Roots. Its golden summer days, cold starry nights, autumn blaze and still winter white are now recreated in the colours and spirit of Roots Algonquin.
Proven in the canoe, on the portage and by the campfire, Roots Algonquin brings a special feeling to work, school and play.

Roots Algonquin...capture the spirit of the Canadian North.

a canoe as often as possible; and Michael Budman and Don Green, also non-Canadian Canada-lovers, have put in some serious paddling time.

You see them canoeing together in old photos: they take to one in the 1997 CBC *Life and Times* documentary made about them; and an editorial cartoon in a 1991 *Financial Times* story about the first stretch of financial rapids in the company's history (cutline: "White Water for Roots") shows the two of them cheerfully navigating a canoe through churning waters. Whatever the cartoon may tell us about Roots' rocky fiscal circumstances — 1991 was the only year in the company's history in which profits weren't made — it assures us that smooth waters lay ahead. For those with the physical and spiritual skills to control it, the canoe bobs like a cork on the surface of adversity. And didn't somebody say that being Canadian meant knowing how to boink in one?

In the Roots story, there are certain narrative refrains
you just can't escape — they stick like a chenille crest. Things
like Mike and Don's Motown boyhood, the lowly negative-
heel beginnings, the shoemaking Kowalewskis, the selling
of team spirit, the recruitment of celebrity and, possibly the
most persistent and unlikely, Camp Tamakwa. Tamakwa is
to Roots what Gordie Howe is to Gretzky or the blues are to
the Rolling Stones. Elementary.

Indeed, when their first contact with Algonquin
Park — Tamakwa's setting — is recounted, the tone is downright mystical. "What struck me that day," said Green in
1990, "was the entrance to Algonquin Park. This magical,
enchanted gate where you stop your car and a warden asks
you a few questions about camping. It's a beautiful stone
structure, and at the time they had some wooden bears on
top of some posts. You pass through the gate and you're in
the Park."

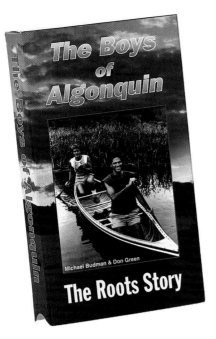

The Boys of Algonquin,
a long way from Motown.

Tamakwa camp poster, 1966, the
year Budman was canoe instructor.
That's him framed in birch.

And once you're there, and the spell has taken, you never fully return. Like the children who find doorways to enchanted kingdoms, the future emperors of Roots were captivated for life: together, they've spent a half-century of summers at Algonquin, and both have built summer properties in the heart of the magic kingdom. Now their children attend camp there — at (take a wild guess) Tamakwa. Sound a little Disney? Maybe.

A 1997 TV company profile, broadcast nationally on the CBC, is not only called *The Boys of Algonquin* — and thus unlikely to be confused with, say, the Perry Ellis or Tommy Hilfiger stories — but the videocassette box prominently features a typically cheerful and hale Budman and Green paddling a canoe through the sparkling waters of their beloved Canadian camp country. In fact, the entire story is framed by the "enchanted" archway of Algonquin — it's what gives narrative rhyme and reason, what makes it cohere as a story. Camp plays a key role in the mythical prehistory of Roots, a role that Budman and Green have promoted almost as aggressively as they have the clothing that Tamakwa profoundly inspired.

The facts are fairly straightforward: during the early Sixties, Budman and Green spent the summers of their formative years at Tamakwa, a private camp named in the faux-Indian fashion of so many North American kiddie camps. While Green and Budman didn't actually meet at Tamakwa (that happened in high school in Detroit in 1962), their discovery of the shared experience there marks

a significant intersection in our story. The revelation that they were ex-Tamakwans sparked the friendship that built Roots and endures to this day, and it seems to be the magnetic core that holds the whole thing together. No Tamakwa, no Greenbud, and most definitely no Roots. That's why, if you want to understand the roots of Roots, you've got to go back to camp.

That's right: *back*. For even if we've never actually been to a Tamakwa-style, cabins-and-campfires summer camp, most of us have been there in the virtual sense. We have an idea of places like it: the gawky kids dressed in camp sweatshirts and sneakers; the canoe-tipping and other good-natured pranks; the hikes, sleepovers and scary moon-lit stories; the tearful partings at idling station wagons — even the alluring smells of pancakes and bacon floating from the mess hall on the hill. For this is the camp foisted on all our imaginations from the heyday of camp culture in the Fifties and Sixties.

Tamaka Rootsified.

Roots Tamakwafied.

An idyllic community of wooden cabins clustered around the gently lapping shores of Spirit Lake, Tamakwa was as postcard-perfect a summer camp as any urban child of the baby boom might imagine. Catering largely to comfortably indulged children, Tamakwa was the annual destination of a far-flung group of city kids from both sides of the border. (Among Budman and Green's camp contemporaries were at least two other lifelong friends and eventual collaborators: future *Saturday Night Live* regular Gilda Radner and movie director Mike Binder, who would eventually make *Indian Summer*, a movie about a Tamakwa reunion deeply influenced by Budman, Green and Binder's shared camp experiences.)

Imagine the impression this rustic jewel in the Canadian wilderness, all emerald green, shimmering blue and electric sunsets, must have had on two Jewish kids from Motown. They must have felt as if they'd stepped directly into the pages of *National Geographic*, or maybe a Canadianized Disneyland: "Algonquin Park," Budman once said, "is an enchanted forest and a magnificent place."

Among the elemental forces at Tamakwa was one Lou Handler. An avuncular middle-aged nature junkie who recruited citified campers by travelling around with Tamakwa home movies, Handler was a former boxer (and violinist, and naturalist and photographer), whose fondness for flannel and khaki was matched only by his shaman-like knowledge of the Algonquin wilderness. Known to all as "Unca Lou" — as though the kids were Huey, Dewey and Louie — he was

Lou Handler at Tamakwa:

the wellspring of team spirit.

rumoured to possess almost mystical tracking skills, and made it nothing short of a personal crusade to make sure that the Tamakwa experience was an indelible one for every child fortunate enough to go there. We're talking father figure of the first order.

Handler's partner at the camp – another first-order father figure – was Omer Stringer, a near-legendary

Boys, dogs, and Omer Stringer in his classic Beaver Canoe, 1983.

above **The quintessential Roots pitch, with home fires burning.** *right* **If this looks familiar, you've probably been to Budman's cabin.** *below* **If the headdress looks familiar, you've probably seen the 'Canadian Pacific' jacket. The personal becomes product.**

Algonquin guide and canoeist with an equal talent for making profound impressions on young minds. In 1995, a full forty years after his first Tamakwa summer and seven years after Stringer's death, Budman posed with Stringer's 1936-issue knapsack for a newspaper column called "My Favourite Object."

Offered by Stringer as a gift to Green and Budman following a 1983 canoe trek, the canvas knapsack has a totemic power for the former Tamakwans. Moreover, it would provide the design for at least one of Roots' lines of knapsacks — another case of the company's tendency to allow professional strategy to be defined by personal experience. In the Eighties, in fact, Roots would help Stringer open his own high-quality leisurewear business, the short-lived but unmistakeably Tamakwafied "Beaver Canoe."

Along with Stringer, Unca Lou looms large in Michael Budman and Don Green's passion for camp life — not just as fond childhood memory but as a professional inspiration. Indeed, if anyone can be called the spiritual father of Roots, it's Lou Handler. His influence is felt in everything from the rugged, all-weather sturdiness of the typical Roots garment to the core philosphy of the company's retail enterprise: you join a team when you wear Roots, a team forged in tandem notions of fun and order. As Michael Posner noted in 1993, even the corporate structure of Roots seems strangely camplike and Handlerian: "Don and Michael — everyone calls them Don and Michael — are co-directors, *sans* the whistles but always enthusiastic, energetic, focused

Beaver Canoe crest:
ready-to-wear Canadian.

The Roots knapsack began
as a paean to Omer Stringer's
1936 original.

BACK TO OUR ROOTS CANAD

on positives. . . . Stressing the ideology of teamwork, they coax and prod some 500 campers — seamstresses, shoemakers, store clerks — pushing them to set goals, work hard, achieve personal bests."

But as indelible as the camp lifestyle was, Green and Budman were every bit as smitten by what was worn there, and to this day the archetypal Roots garment evokes echoes of classical sport, school and camp apparel: the uniform of good times past. Yet unlike, say, boot camp or prison, where the wearing of a uniform implies a forced suppression by a greater authority, the Roots notion of teamwear is loose, open to individualization, and fairly reeking of good, clean, healthy fun. In other words, when you join the Roots team, you don't hand in your distinctiveness. The uniform ties you to a sense of collective effort and shared responsibility, where the fun to be had is increased by the number of people willing to play. And what do you get from it? Well, as anybody who's been to camp knows, you get out of it what you put into it.

At least that's the theory, and it's a theory that would drive the eventual success and distinction of Roots. Demonstrating the productively arrested development that would become a generational hallmark of the baby boom, Budman and Green not only shared a childhood experience at Tamakwa that adulthood could not dim, they discovered the Holy Grail of boomercult: the secret of making a living — and a damned good one — out of not growing up. Essentially, the secret of Roots' Tamakwa influence is this: Michael and

Team Tamakwa: college goes camp at Roots.

opposite Indoor activities: taking Roots from the bush to the mall.

1960

Voyageur 4

top **Don, big brother Richard,
and their mother Bethea, camp
Michigama, Michigan, 1959.**

above **Michael Budman, front row
third from left, Tamakwa, 1960.
In time he'd be standing.**

below **Michael, back row centre,
Tamakwa section head,
1968. In charge at last.**

Don were not only able to keep the Tamakwa campfire burning well into their own middle age, but they also gave everyone else an opportunity to buy a piece of the camp experience. To feel like part of a team. Kind of like what Lou Handler did, minus the lake and canoes. But who needs lakes and canoes as long as you can get the sweatshirt, the *skin* of the camp experience.

That camp culture, a fairly rarefied phenomenon in the century's early decades, was a firmly established element of apple-pie Americana by the mid-Fifties, when Lou Handler schlepped his portable projector around to woo kids into summering amid the Kodachrome splendours of Tamakwa. Which makes perfect sense, when one considers the general trajectory of kidthink over the course of the American century. While wilderness camps were definitely up and running by the Depression — when Unca Lou and Omer Stringer blew Tamakwa's first reveille, when the period's economic anxieties led to a surge in anticipatory anxiety about the fate of America's — as one famous movie called it — "Dead End Kids," it took the particular socio-political proclivities of the Eisenhower years to bring camp smack into the postwar mainstream.

For concurrent with the rise in postwar prosperity and leisure — not to mention the widespread residential exodus to the vast anti-wilderness of the suburbs — was a widespread cultural determination to keep the greatest demographic bulge of the century firmly under control. Significantly, the rise of camp culture coincides with the

rise in panic over teenage delinquency, the fear that the country might be taken over by hordes of teens and pre-teens. Naturally, the rise of rock music, with its perceived unleashing of adolescent libido, didn't help. Indeed, after the bomb and Communism, teen phobia was the great bugaboo of the Eisenhower years. This, after all, was the time of rock & roll record burnings, James Dean, and hor-monal hysteria movies like *I Was a Teenage Werewolf*.

Camps such as Tamakwa, not unlike paternalistic

Don Green, back row centre, a Tamakwa counsellor. By 1969, camp had become a looser outfit.

Roots camp:
a good place to go.
Guaranteed good weather,
great cabin mates, and no bugs.

situation comedies such as *Leave It to Beaver* and *Father Knows Best*, or a book like Dr. Spock's, filled a key reassurance function for parents: kids sent back to the land to be instilled with a good, old-fashioned sense of order and responsibility were more likely to navigate the rocky shoals of adolescence without recourse to switchblades, Brylcreem or the devil's music. Or so Mom and Dad hoped.

It didn't work, of course. After all, Elvis Presley never went to camp but Lee Harvey Oswald did. Imagine how history might have changed if things had been reversed. Indeed, for all the generational paternalism at work in the Fifties – possibly *because* of it – the kids of the Sixties turned out to be the most obstreperously unruly the century had seen. Rock music became the ubiquitous music force in mass culture, fear of Communism and nuclear attack mutated into something resembling a generational civil war, and even the message of camp culture changed. Certainly the popular impression of camp altered profoundly, from the tranquil postcard idealism of the Fifties to an object of widespread satire a decade later: Allan Sherman's "Camp Granada" ("Hello Mudda, Hello Fadda . . .") reflected this just about when *Mad* magazine began poking ongoing fun at the conformist pretensions of summer camp. In 1965, the transformation of camp from ideal to punchline seemed complete with the appearance of a network TV series called *Camp Runamuck*. The premise of this weekly comedy – like that of such movies as *Meatballs* a decade later – was to make fun of camp life for precisely the reasons it was held to

be valuable a decade earlier. By the mid-Sixties, of course, it was easy to laugh: the failure of summer camps to prevent widespread social upheaval was as obvious as the radio domination by rock or the riots soon to explode on the streets of Detroit, that paradigm of orderly Eisenhowerian industry.

For the purposes of our story, however, two facts concerning the changing cultural profile of camp need be borne in mind. First, neither Budman nor Green soured on camp life. Indeed, one might argue that they've never really left. Second, like for a lot of counterculturally inclined boomers, the camp experience proved a defining one in their emerging worldview. Avowed environmentalists and spokespersons for the preservation of Canadian nature and wildlife, Budman and Green had the seeds of their own eco-consciousness sown beneath the verdant branches of Tamakwa. Significantly, as well as giving them an ideal to market a retail lifestyle on, Tamakwa gave them something to rebel against, or more accurately, to stand for, a key distinguishing feature of the Roots retail crusade.

Finally, the impression of Canada that these Motown kids derived from those perfect Tamakwa summers cannot be overstressed. Indeed, it can be reasonably suggested that, if it weren't for Tamakwa, it's highly unlikely that either Budman or Green would have made Canada home. Certainly it was Tamakwa that gave them the idea of Canada that they would later transform into an intensely attractive element of their pitch: a country as rugged and unspoiled as it is fragile and beautiful, vast expanses of green wilderness

Tree hugging: Tamakwa crest, in chenille, the precursor of the Roots beaver.

Canada in chenille: mythic, sturdy, impeccably stitched.

Jeanette MacDonald and Nelson Eddy in *Rose Marie*, 1936: even then, Canada was a romantic American myth. Still, not a bad fantasy, is it?

intersected by pure and sparkling water systems. The Roots logo, in fact, is a variation on the Tamakwa crest, which pictures a beaver in silhouette chomping on a tree. In the Roots worldview, Canada is a blessed place for work and play, and a fine one never to have to grow up in.

Of course it's a mythical Canada, no more the country at large than Monument Valley is the U.S. or the Yorkshire moors England, but no less enticing for that. On the contrary, perhaps it is *more* enticing for being mythical. Ironically, for all the Roots story reveals about the urbanization of Canadian culture during the Seventies, the image of Canada marketed through Roots products hearkens back to the Canuck kitsch culture of the early twentieth century, when people around the world were flocking to silent movies that sold Canada as an exotic untamed wilderness populated by moose, snow, mountains and Mounties.

Roots clothes, in their sturdy, all-natural traditionalism, imply a country of canoes, cottages and campfires, where everyone spends a lot of time outdoors because outdoors is nowhere better than it is here. It's a vision of Canada largely rejected by post-Expo 67 Canadians determined to appear as downtown as anyone else, yet it's turned out to have far more resonance than anyone — including Budman and Green — had imagined.

Arguably, it took a couple of kids from Detroit — the other side of the border from which Canadians watch America — to see the viability of this mystical Canada as a lifestyle and marketing principle. Not to mention seeing it as Canada itself, for it's hard to imagine any home-grown Canadians other than those in the tourist industry buying the "Hinterland Who's Who" myth after the cultural upheavals of Expo, Trudeau, the October Crisis and — just a year before Roots opened shop — the Commie-kicking Canada-Soviet hockey series of 1972. At that point, those in a position to determine the national self-image were trying to put as much distance between the "new" Canada (hip, urban, contemporary) and the "old" (traditional, rural, timeless) as possible. But, as it turned out, old was an image Canadians were hungry to buy. Literally so. The mythic resonance of postcard Canada — the unspoiled hinterland — may touch a deeper national chord than many of us realize. So deep, in fact, it would seem to transcend nation itself: one of the truly arresting things about Roots' success is not just that it successfully sold an idea of Canada to Canadians, but that it would eventually sell it to much of the world. And to think it all started at camp, with Unca Lou, Omer Stringer, and a few lessons in how to handle a canoe.

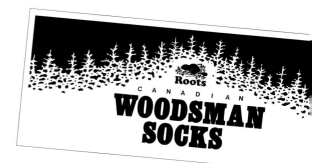

For the logger in us all.

Heather Cooper's 'Woodland Series': canoe, beaver, lakes and lapping shores. Wish you were there?

TAMAKWA — THE MOVIE

DON GREEN
VINCENT SPANO

KEVIN POLLACK

MIKE BINDER

MICHAEL BUDMAN "UNCA LOU" HANDLER

Roots

Indian Summer postcard:
Roots meets Disney, a match made
in Maple Leaf heaven. Just ask the
Mounties, now licenced by Mickey.

Viewed outside of Roots culture, the 1993 Disney-produced movie *Indian Summer*, written and directed by former Tamakwan Mike Binder, and featuring more conspicuously displayed Roots clothing than a warehouse sale, is unspectacular stuff. Another one of those *Big Chill*-like baby boom reunion movies, it's about a group of old friends — including two brothers who've become successful casualwear clothing entrepreneurs — summoned to their childhood camp for a week-long orgy of regret, regression and shamelessly tearful nostalgia. A typical line: "I wanna be a kid again. I was good at being a kid, wasn't I?"

Viewed *as* Roots culture, however, the movie is something else: the big-screen version of the Tamakwa myth, a multi-million-dollar, feature-length commercial for camp culture, Roots-style.

Set at Tamakwa and shot in Algonquin Park, the movie is a thinly fictionalized but highly sentimentalized expression of Budman, Green and Binder's Tamakwa memories, right down to the goofy pranks, camp-specific lingo and, of course, Unca Lou. Played in the movie by Alan Arkin, Lou is nothing less than a fount of paternal wisdom and comforting, flannel-covered constancy. It's he who has invited this group from (as he calls it) "the golden age" for the reunion that will re-invigorate their lives; and it is he who engineers their redemption from the corrupting influences of childhood.

It's a movie in which people seem to say "I'm sorry" a whole lot and, in the hermetic moral universe *Indian Summer* creates, there would seem to be nothing that simply

Scene from *Indian Summer*: Alan Arkin as 'Unca Lou,' with Kevin Pollack and Bill Paxton as superannuated campers. As one Roots ad none too subtly put it: "You've seen the movie, now wear the clothes!"

being sorry can't solve. Like a morning dip in a northern lake, everything, even the darkest psychic stain, is washed clean and purified by apology.

Needless to say, the Disney connection here is evocative, for no corporation has more successfully marketed fantasy and the appeal of never growing up than the one that brought *Indian Summer* to the screen. Not only that, no entertainment company has capitalized more successfully on the particular state of arrested childhood that is the baby boom's — and no clothing company has mainlined the particular contradictions of the baby boom than Roots, by packaging lifestyle fantasy in the form of kids' clothing for reluctant grownups. It's fantasy, and fantasy provides temporary relief from contradiction, a brief but powerful haven from the messiness of the present: a return to a "golden age," when everything was simple and tidy, when someone told you when to get up and what to do, and when simply saying sorry solved everything.

Or, as Unca Lou tells the gang at one point: "The only reason you like this place is because the rest of your lives are so boring."

Roll with it: anonymous hipsters hang in Yorkville, Toronto, in the Sixties.

Over the Counter Culture

Roots, Rock, Retail

"Most important of all, the counterculture is said to have worked a revolution through lifestyle rather than politics, a genuine subversion of the status quo through pleasure rather than power."

    — *Thomas Frank, The Conquest of Cool*

"There's change for ya. What used to be an anthem against people like the bank is now a jingle for the bank. And it's a nice jingle, too. If you listen carefully, you can actually hear the sound of Woody Guthrie spinning in his grave."

    — *Rick Mercer, on Bob Dylan and The Bank of Montreal,* This Hour Has 22 Minutes

Unlike the Sixties, when it seemed that everyone under twenty-five — save maybe Frank Sinatra, Jr. — was onside with the revolt against the Establishment, these days it's a rare happening that can close the ranks of the baby boom. After all, when the "boomer" designation applies to everyone

**Make dams, not war:
Roots takes a stand.**

Bob Dylan in "Don't Look Back,"
1967. By 1995 he was singing for
the suits. Invest early, retire late.

from Bill Clinton and Michael Eisner to Keith Richards and
Archie, it may be time to admit that demographics may be
the only thing — besides suspicion of youth and a dubious
fondness for "classic rock" — that holds the century's once-
greatest youth movement together.

Yet, every once in a while, something resuscitates
that old collective spirit, or some ghostly echo of it — and
it's usually something that threatens the heavily guarded

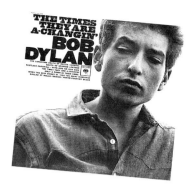

mythology of boomerism itself. Like war veterans or former fraternity brothers reuniting in retro-outrage over some slight to their collective memory, no matter how dispersed the boomers may be, they can pull together like southbound geese when there's a threat to the past.

Take the Bob Dylan bank commercial. In 1995, with a boomer in the White House and the *Forrest Gump* classic rock soundtrack everywhere, an uproar occurred over the defilement of a sacred boomer hymn – Dylan's generational anthem "The Times They Are A-Changin'." The Canadian mega-Bank of Montreal, seeking to recast its image in a more plebe-friendly and streetwise light – "Can a bank change?" it asked – used Dylan's words and music in a promotional campaign to convince customers how much it really – no, *really* – cared. To this end, it deputized not only Dylan and his song, but possibly the most famous visual trope from the seminal 1967 Dylan rock-doc *Don't Look Back*. Except this time, instead of a sinewy young Bobby flashing lyric-covered cue cards in a garbage-strewn alley, the commercials pictured corporate-issue "average Canadians" holding up cards expressing their decidedly non-revolution-ary hopes, dreams, fears and aspirations. The kind of con-cerns, in other words, better alleviated by sound investment than social revolution. "For some reason lost on me," noted comedian Rick Mercer at the time, "it makes sense for a bank to promote itself by showing us endless pictures of people that a bank wouldn't cross the road to spit on, let alone lend money to."

The selling of subversion: rock retails revolution.

Love-in in Hamilton, 1967: all are loved except spoiled, middle-class "Yorkville Hippies." The fashion mannequin points to the future.

opposite Green, Budman, dogs and collars, 1975: Calling all hippies, including the Yorkville variety.

Bob's bank campaign backfired big-time with the very demographic it sought to seduce: outrage clogged e-mail circuits across the country, and the commercial became synonymous not with the attainment of dreams through better banking, but the bankruptcy of the dream itself. A long, strange trip indeed.

Yet, as it was when Nike's use of The Beatles' "Revolution" met with boomer horror in 1989, and Jane Fonda cashed in whatever residual countercultural cred she had when she said "I do" to Ted Turner, a vein of denial ran through the righteousness. (Moreover, it's worth noting that the rancour was overwhelmingly directed not at Bob but at the bank, despite the obvious fact that Dylan must have profited handsomely from the song's use. But that's the beauty of myth: facts are irrelevant.)

The thing was, the song's sentiment was right — just a little too right, and its use in a bank commercial proved the point. The times they *had* a-changed, just not in a way

anyone wished to admit. It wasn't so much that the Bank of Montreal was betraying Dylan's song as draining it of its "revolutionary" integrity, thereby exposing one of boomer-cult's greatest fears: sellout. It was one thing to embrace a movie like *Forrest Gump*, which valorizes denial and a mushy imperviousness to reality, but it was another to recruit Saint Bob – who'd soon be singing "Blowin' in the Wind" for the Pope – to peddle registered retirement savings plans. It was one thing to know that the dream wasn't real, quite another to *advertise* the fact.

As a company, Roots has wisely never stood on its countercultural integrity alone, though its success cannot help but be viewed on the post-Sixties trajectory of the baby boom. Not only are its founders and owners unabashed boomers with their feet in the Fifties, hearts in the Sixties and heads in the present; the company can be fully understood only in the context of the post-countercultural moment in which it was born.

"*Post*-countercultural" needs stressing here, as by the time Budman and Green opened their first shop on Toronto's Yonge Street in 1973, not only were the Sixties most decidedly over, but the true legacy of the era was only beginning to become apparent. The seeds had been sown, but the harvest wasn't quite what some had imagined. A funny thing had happened on the way to the Seventies. Thomas Frank wrote in *The Conquest of Cool*, a study of the counterculture's abiding legacy of "hip consumerism":

Hippie of indeterminate gender makes love wagon even lovelier, possibly inspired by Led Zeppelin lyrics on the back of the t-shirt, " . . . and as we wind on down the road . . ."

The countercultural style has become a permanent fixture on the American scene, impervious to the angriest assaults of cultural and political conservatives, because it so conveniently and efficiently transforms the myriad petty tyrannies of economic life — all the complaints about conformity, oppression, bureaucracy, meaninglessness, and the disappearance of individualism that became virtually a national obsession during the 1950s — into rationales for consuming.

Thus, the failure of the flower-powered people's revolution notwithstanding — in '73 Nixon was still President, Hoover still ran the FBI, and the comeback kid of the year was Frankie "Ol' Blue Eyes" Sinatra — the impact of the Sixties was nevertheless both profound and ubiquitous, and nowhere more visibly so than in the realms of fashion and culture. If anything had been revolutionized, it wasn't politics or economics, but attitude and style, the twin pillars of contemporary consumerism.

Buy buy love: countercultural marketing in embryo.

Rock music, once a freakish entertainment sideshow, had firmly taken its place as the pre-eminent commercial musical force: it was the dawn of arena rock and "superstars"; the once-underground *Rolling Stone* had moved from Frisco to New York and gone successfully upmarket; and even network TV had finally yielded to the call of the wild. *In Concert* and *Don Kirshner's Rock Concert* became late-night staples, and talk show couches were suddenly full of the unkempt likes of John Lennon, Lou Reed and David Bowie. *The Partridge Family*, a sitcom about a suburban family that also happens to be a rock band, was one of the year's top-rated shows.

Hollywood, which during the Sixties had tried to put as many bloated musical turkeys (*Paint Your Wagon, Darling Lili, Doctor Dolittle*) as it could between itself and a counterculture it clearly found flummoxing, had finally had its corporate head yanked from the sand. Awakening belatedly to the boardroom-rattling success of such hip, low-budget smashes as *Bonnie and Clyde, Easy Rider, M*A*S*H* and *Five Easy Pieces*, the studios began bankrolling home-grown "art" movies: *Badlands, Brewster McCloud, Mean Streets, Scarecrow, Two Lane Blacktop.* Invariably made by young film-school grads – with decidedly unAmerican names like Bogdanovich, Coppola, Scorsese, De Palma, Polanski and Cassavetes – who had spent the Sixties listening to rock, watching Euro-art movies and worrying about the sorry state of the union, the movies of the early Seventies were centralizing the fringe: movies weren't considered

worthwhile unless they were "relevant," and stories critical of the so-called "establishment," whether direct or oblique, socialist or reactionary, sincere or bogus, were suddenly springing up everywhere, like sideburns. At least in terms of popular culture, the real impact of the Sixties hadn't begun to be felt until the early Seventies. It was the era of *The Mod Squad* and *All in the Family*'s Meathead, when Mel Tormé sang Donovan, and Prime Minister Pierre Trudeau looked like a refugee from a Dennis Hopper movie.

Maggie and Pierre in disguise as folks at the Perth Summer Festival, July 16, 1971.

77

White suit, pant suit, tacky tie and peace sign: the dawn of the decade taste forgot. By 1970, countercultural fashion was going uptown . . .

Fashionwise, things were the sartorial equivalent of nuclear meltdown, and the most dramatic public evidence of how the times had a-changed. Elements of radical youth fashion — long hair, bell bottoms, sideburns, scarves, bright colours — collided and co-existed with the business suit and formal evening wear, and even the most established department stores began to open boutique corners for hipper suburban shoppers. (Some Canadian cities even boasted a countercultural fashion chain whose name said it all about the peculiar upheavals of the age: "The Establishment.") Needless to say it was the era of the leisure suit, that unholy melding of cool and conservative that could have seemed to be a good and necessary idea only in the early Seventies. In retail terms anyway, the line between radical and mainstream was trampled by the stampeding platformed hordes, and the real legacy of the Sixties Youth Rebellion was laid bare: with a well-aimed appeal to people's hipper instincts, you could sell anything. Even a shoe with a negative heel.

. . . leaving places like Yorkville blowin' in the wind. Until the developers arrive, that is.

Don Green, Denyse Tremblay, David Suzuki and Stein Eriksen at Roots-sponsored fundraiser for Suzuki's environmental foundation, 1996.

Another kind of saving: consumerism with a conscience.

It's probably safe to say that Roots wouldn't have taken off in quite the same way at any time but the early Seventies. Naturally, part of this is attributable to the anything-goes nature of the day's fashion tastes. But there was more to it than that.

There was the fact that the two future clothing moguls were most definitely post-countercultural boomers. Not radicals or underground press barons, but a couple of kids from upper middle-class Detroit who couldn't help but be influenced, and deeply so, by what had transpired during the decade of their coming of age. Consider the following:

Roots was made possible by dropping out. Michael Budman had dropped out of teaching in the search for something more meaningful — a perfectly reasonable thing for a young adult to do in those days. Don Green had dropped out of university. Fresh from two years of bumming in Jamaica at the time he reunited with high school buddy Budman in Toronto, Green was a classic post-hippie

Preserving the rainforest:
Brazilian natives, Green
and a pensive Sting, 1989.

The message is the message.

entrepreneur. Fed up with conventional career-building, he decided to do it his way.

Roots might have been a yogurt store. Actually yogurt was only one of the post-countercultural product options The Boys considered before they stumbled over the negative heel. Futons were another, as were flowers. There is no evidence that they pondered a blacklight franchise, but it would come as no surprise. Besides, their shoe was the ultimate product of the moment: non conformist, hand-crafted, and imbued with the spiritual experience of standing with one's feet in the sand. A shoe for former sandal wearers and wannabes who never really had the nerve. Or possibly the toes.

Since the beginning, Roots has been down with boomer-friendly liberal causes: environmentalism, literacy, anti-racism. Like so many Sixties-steeped holdovers, the company has maintained the crusading patina of the decade, but never in a manner (and never for issues) that

APEC finance ministers Peter Costello (Australia), Dato Haji Selamat Haji Bin Munap (Brunei) and Paul Martin (Canada), enjoy their spiffy new Roots duds, Kananaskis, Alberta, 1998: some job we got, huh?

might intimidate consumers or generate controversy. It has not, for example, crossed the fine line between supporting causes and political endorsement, as evidenced by indiscriminate freebie handouts to everyone from Bill Clinton and Jean Chrétien to Ontario Premier Mike Harris and Prince Charles — in the Roots marketing universe, politics would seem to be just another irresistible photo opportunity. In post-countercultural marketing, activism is good only if it is worthy and non-divisive. For the implied message of all post-countercultural marketing phenomena is: having done so well by the system, how bad can the system be?

Like many of the day's products, the negative-heel shoe both prided and sold itself on laid-back non conformity. Not just in design and intimations of healthier living, but in marketing. Early Roots ads emphasized craft over style, appealing to people who liked to think their interests transcended fashion and status. Moreover, the ads often emphasized the craft of the shoemaking Kowalewski family, another appeal to the back-to-tradition sentiment of the post-hippie pitch.

Most important, Roots was not shy when it came to calling its products "revolutionary," possibly the most potent marketing staple of the era — by the late Sixties, everything from cola to compact cars suddenly became an agent of insurrection. In doing so, the company was appealing to the residual sense of upheaval left over from the Sixties, displaced from politics to marketing. To wear this "revolutionary" shoe, in other words, was to indulge one's nonconformity, but in the apolitical and consumerist sense that had overtaken the word by the time Roots opened its doors in 1973. For by that time, the very idea of revolution had transmogrified into what would symbolize the real legacy of the Sixties: while the world hadn't been overthrown by the youth-generated upheavals of the era, the language of upheaval, the attitude of rebellion, was here to stay. Strangely, considering the ramparts-storming rhetoric of the Sixties, one of the most profound and lasting beneficiaries of the era's revolutionary temper would be capitalism. Revolution had become consumer choice, rebellion an attitude, and nonconformity boiled down to the way your heels happened to tilt.

THE CANADIAN SPORT ROOT

Where good shoes go when they die.

Clearly, by 1973, the times they had a-changed, but the song, as the quintessential Seventies band Led Zeppelin put it that year, remained the same.

Family value:
Don, Michael and rootlets
show the company colours.

My Pleasure is My Business

Roots Gets Personal

"As medical science has extended the life expectancy of Canadians, our quality of life has suffered. What many call the "greed of the Eighties" has left our fundamental needs at risk. Our personal health and fitness, our environment – even our family traditions are in peril."

   – Roots Report!!, *in-store newsletter, 1989*

"If we like it, if it's something we'd wear, then we're pretty sure that other people will too."

   – *Don Green*

"It's business, Sonny," mafia *consigliere* Tom Hagen reminds the hotheaded, vengeance-hungry mafia heir Sonny Corleone in the movie *The Godfather*. "It's *not* personal."

   Good thing Michael Budman and Don Green didn't seek Hagen's advice, for this is one concept they'd have a hard time with. For these two, there is no clear border where business ends and families, hobbies and friendship begins. In fact, it's their indistinguishability that's the point. In Roots

85

corporate culture, what's business is personal and what's personal is business, and woe betide anyone in the system who sees them as separate. Not that they'd find themselves sleeping with fishes, but if they don't get it, they don't get Roots.

This investment of the personal manifests itself in just about every aspect of the company's public personality. First there is the mythic role of the Budman-Green friendship in the company's history: "We got together the same year the Stones did," Budman has said. The yin and yang of their personalities is as if a single genetic entity were divided into separate but symbiotic bodies. After all, these guys not only work together in the close and pressurized proximity that can tax even the strongest marriages, but they take summer vacations together, a stone's skip from where they both went to camp. (Not only that, they genuinely seem to *like* each other.)

The relationship has been seamlessly integrated into company history from the beginning, both as an irresistible journalistic hook and as an advertising strategy: Green and Budman have been company clothing models intermittently since the beginning, and a persistent theme in Roots advertising is the personal nature of the products bearing the company name. Emphasizing the role of family, loyalty and friendship in the production of quality goods — guess it boils down to that thing called caring — Roots has made the promotion of the personal a key to its ongoing market strategy.

Moreover, it's not hogwash in the way that one suspects that the guy in the Wendy's hamburger commercials probably really doesn't eat at Wendy's much or that Kathie Lee Gifford

Let's get personal: Budman and
Green promote the product.

All in the family:
right **Bloor Street store,
Toronto, designed by Budman's
wife, architect Diane Bald.**

below **Michael Budman and
brother Jim, who played a large part
in designing Roots' New York store.**

likely doesn't buy her clothes at Zeller's. Roots' owners are
always dressed in head-to-toe Roots, giving credence to the
claim that no item of clothing gets the conspicuously displayed Roots logo unless Budman or Green — or their wives
or children — have worn and endorsed it first.

But it's also evident in the role the owners' families
have played, and continue to play, in the Roots saga. In
the beginning, at the dawn of the so-called "Me Decade,"
both men's successful and savvy businessman fathers were
instrumental in offering advice, guidance and the occasional
cash advance to the fledgling company. (The Kowalewskis,
the shoemaking family who've overseen the production of
the company's leatherwear for a quarter-century, are another
manifestation of the company's family-oriented public
profile.)

Budman's wife, architect Diane Bald, designs Roots
retail environments; Green's spouse, Denyse Tremblay,

was hired as one of the company's first salespersons and continues to play a key role in determining company policy. Moreover, none of the women's clothing finds its way to the racks until Bald or Tremblay have worn it — and liked it.

Lately, this what's-good-for-the-family approach has gone multi-generational. Not only did Roots not begin producing clothing for children until the owners began producing children, but the kids are the primary fashion consultants when it comes to non-adult apparel: it was Budman's preteen son Matthew who gave the thumbs up to the blockbusting polar fleece poorboy.

There's more to this strategy than good, homespun public relations. When it pays off, it really pays off.

Come to think of it, if it didn't pay off, it would seem downright arrogant. For what's really at work here is the vintage Me Decade assumption that what's good for our family must be good for everybody's. Moreover, they couldn't be more upfront about it. Ask either of them how an item of clothing comes to be incorporated in the Roots line, and you're not likely to hear about market research or focus groups. Company policy is based on an intensely unscientific combination of observation, clairvoyance and gut instinct. (Indeed, "What's your gut?" is Greenbudese for "What do you think?") If it feels and looks good, if it meshes with their image and lifestyle — and if it can be affordably produced by the available manufacturing facilities — then it's probably going to find its way onto a knotty-pine Roots shelf near you. Not out of arrogance but in the belief that, if they like it,

above **Father and Sons: Irwin Green with sons Don and Richard.**

below **Michael, wife Diane Bald and son Matthew afloat in Roots.**

bottom **The son also rises: Matthew, the kid who fast-tracked the poorboy cap also inspired the leather college jacket. Here he models the former, surrounded by his family, at the Nagano opening ceremonies.**

roots kids

Roots kids:

The line was born

only when there were little

Budmans and Greens to wear it.

their customers will too. This, along with rigorous supervision of quality standards and hands-on production methods, may be all one really needs to know to understand Roots: Budman and Green see themselves as surrogates for the consumer at large, as people whose values have a certain universal appeal and application.

In a way, this process represents yet another manifestation of the strength, nearly three decades later, of the company's post-countercultural roots, as it maintains a sense of the hands-on, mom-and-pop personal touch even now that it's an internationally successful clothing company. In this regard, "Mike and Don's" is the garment version of "Ben and Jerry's" ice cream: an empire built on good old-fashioned small-business principles.

Viewed straight on, the notion that these guys have their pulse on the taste of the lumpen is absurd. These, after all, are self-made millionaires with multiple residences, celebrity friends, personal trainers, and sufficient accumulated air miles to charter the Concorde for every trip to the 7-Eleven — so just how "universal" can their take be? Quite, as it turns out, and for a simple reason: in intuiting that their customers, in age and inclination if not income, are very much like the Roots boys themselves, Budman and Green sell their reality as fantasy fodder for the similarly inclined. Not people who have Green and Budman's lifestyle, but people who wish they did, and who can purchase a piece of that wish at Roots. For what Roots clothes symbolize is the triumph of leisure and the never-ending good time, a perma-

nent vacation from grownup worry and care that, as fantasy — particularly as *boomer* fantasy — gains potency the farther it is from most people's reality. And what is consumerism other than dreams for sale?

It's not the jet set, executive-class, friend-of-the-famous lifestyle that Roots sells — it would probably have long ago gone the way of Robin Leach had it tried. What it sells instead is a fantasy variation of Green and Budman's past: the combined cultures of summer camps, sports teams, fraternities and endless holidays, those close associations of fun-loving boys and girls that exert a nostalgic pull so strong it ultimately doesn't matter if that past was your past or not. Like a lump-in-your-throat greeting card or an eye-moistening long-distance telephone commercial, the clothing appeals to what one wishes had been. For many adults the appeal is downright irresistible: it evokes that perpetual state of childhood and adolescent freedom, before grownup headaches and responsibilities, when the only thing worth worrying about was who was going to bring the ball and what time to be home for supper. For any grownup, it's alluring, but for boomers — possibly the first generation to turn its recollected childhood into a cultural shopping mall — it's particularly enticing. You might call it the Peter Pan Principle of lifestyle marketing: what people are really buying is Neverland, the place where you never have to grow up.

On the waterfront: maybe they have cabin fever — sitting on the dock of the bay.

bottom **roughing it in the bush: Green and Budman alert the wilderness to an important anniversary.**

The last thing you'll read:
Muhammad Ali
punches for Everlast, 1974.

Branding Irony

Or, Pardon Me, Your Roots Are Showing

There's a scene in the quintessential mid-Eighties movie *Back to the Future* — about a lovable time-travelling yuppie who tumbles into the Fifties — that tells you a lot about one of our major fashion phenomena. Stripped to his sporty Calvin Kleins in the presence of the randy young woman who's actually (gasp!) his mother, our time-shifting hero, played by diminutive ex-Edmontonian Michael J. Fox, is puzzled when this Eisenhower-issue young woman (Lea Thompson) starts referring to him as "Calvin."

After all, back in those days, the only reason one wore labels on the outside of one's clothes was to keep track of them and/or not to breach prison regulations. Spotting the name on Fox's skivvies, Thompson just naturally assumes the guy's name is Klein. Fact is, if it were a sweatshirt instead of briefs, she might just as easily have been calling him "Roots."

**Your butt is mine:
the Levi's red tab.**

Hard as it is to believe, not long ago — well after Eisenhower departed for that great golf course in the sky — people rarely let their labels show. Indeed, apart from the Lacoste alligator and maybe a tiny Levi's red tab, if a brand label was visible, it usually meant someone had their shirt on inside-out.

Until the mid-Seventies, most clothing manufacturers — even sport- and leisure-clothing manufacturers — kept their labels discreetly tucked away in what was widely assumed to be their proper place, i.e., on a garment's *inside*. If a garment's brand was visible, it was in the overall appearance of the article: the particular cut of a jean told you if it were made by Lee or Levi's, and the style of a shoe let you know if you were looking at Florsheim or Gucci. Indeed, a sure sign of comic hickdom — à la Minnie Pearl or Red Skelton — was wearing a label, price-tag or other sign of commercial origin on the outside of your clothes. It labelled you a goof.

Significantly, it would soon turn out, conspicuously "branded" clothing would soon become apparent in sport: Muhammad Ali's Everlast boxing shorts, Converse canvas basketball sneakers, Mark Spitz wearing nothing but Olympic medals and a Speedo. Even then labels were far more likely to be conspicuously splashed across sports equipment — Louisville, Spalding, Bauer, Titlelist, etc. — than on clothing. One reason for this might well be because sport clothing was already labelled with team logos, symbols and names. Still, the maker of the garment was considered to be of far less consumer interest than the team wearing it, and the "label" one wore promoted the team, not its outfitter.

This began to change in the early Seventies, right around the time Roots opened its first store; and the forces behind the shift seem to emanate from a number of corners. Indeed, while Roots was making a name for itself with a "revolutionary" shoe, athletic-shoe brand consciousness was just kicking in. Adidas began saturating the market with its name and logo, and not only on shoes and shoe ads. "Adidas" was as likely to be found on T-shirts and gym bags and, for a brief moment, it seemed to be the only label that was. Then everybody else caught on.

Meanwhile, designer consciousness was trickling down to mainstream clothing via the pioneering label marketing of companies such as Pierre Cardin. One of the first "prestige" Euro-designers to go boldly mass market, with labels conspicuously blazing, Cardin recognized the growing middle-class appeal of designer clothing and accessories, hitherto the domain of the wealthy, the beautiful and the otherwise unattainable. Cardin also understood that there was more to it than selling shirts. By producing a range of Pierre Cardin accessories — scents, belts, handbags and eyewear — the company was not just anticipating the snob appeal of lifestyle marketing, which would become a designer commonplace by the mid-Eighties, it was setting the rules by which, nearly three decades later, the business of label marketing is still played.

Being in the business of making the flagrantly non-essential somehow appear essential — perhaps the defining marketing challenge in a consumer society — it should come

Designer planet: hip-hopped Filipino teen in Hilfiger.

Generation Swoosh: Nike likes it.

Hip hop high style: sport and attitude.

as no surprise that eventually designers would shift from selling articles of clothing, or even the prestige of their names, to marketing "lifestyle." Selling clothing, after all, is about goods, whereas the selling of lifestyle indulges what may be the single most important enterprise of postwar consumerism: the marketing of fantasy. Dreams you can buy.

Once we understand this, the relationship between designer marketing and popular music movements of the Seventies becomes clearer. One cannot, for example, divorce

mass-market designer consciousness from that blatantly fashion- and status-conscious musical movement, disco — the movement without which the world might have been denied the spectacle of "designer jeans."

Nor, a few years on, can one separate the mass market explosion of label-branded athleticwear from the late-seventies origin of hip hop. In appropriating team and label-licensed athleticwear as a lower-class, black, anti-fashion statement, hip hop set the course for the phenomenal

Roots meets hip hop halfway:
Rupert Harvey of Messenjah (right),
Kim Campbell (below)
and Puff Daddy (bottom)
get down to Roots.

popularity of designer sports, arguably the defining fashion development of the following two decades.

Disco's heyday may be historically coincidental with Roots' origins, but it's hip hop that helps the rise of Roots make sense. For if hip hop — like disco before it — represents an underclass phenomenon that went gradually mainstream, Roots is a byproduct of the point at which various tributaries feed into wider and deeper channels of culture and commerce. Naturally, nothing makes the passage from fringe to mainstream without dilution and loss of genetic purity, and such is

hip hop's relation to Roots. First and most obvious is the fact that, up until recently anyway, Roots' hardcore clientele hardly consisted of young, black, urban hipsters — who could be whiter than Wayne Gretzky, Elvis Stojko or Prince Charles? But hip hop was ready for Rootsification, being well on its way to the malls and boutiques of the suburban consumer.

Moreover, by the time that hip hop had reached far enough into middle America to mingle with Roots, only some of its cultural origins remained intact. The music held little interest for Roots — far less than the idea of conspicuously labelled athleticwear boldly asserting team and brand loyalty. This in it-self was a dilution, in this case of the wearing of "colours" to semaphore gang affiliation. What Roots' clothing took was the idea of team loyalty, thus effecting a mainstream

Roots posters designed by Bruce Mau: bold, strong, and simple. Sporty, yes, but funky too.

What the players really wear:
Doug Gilmour calls all
players to join the
Roots team. No, *really*.

re-appropriation of the athleticwear that hip hop had appro-priated in the first place. For Roots-wearers, the regalia of teams suggests a virtual community, an association defined less by shared interests and activities than by the act of con-sumption. For the Roots gang, like the street gang, the link remains: in our clothing we proclaim community. We appear to belong.

One feature of Roots label-branding is the rather star-tling absence of a consistent visual representation of the company name: while the brand remains consistent, the logo constantly changes. Bold and chunky or tall and thin, the word "Roots" has rarely appeared in the same manner for much longer than it takes to shift from one marketing season to the next. Needless to say, most of the high-profile athlet-icwear designers cling to a single logo with the graphic tenacity of a tattoo.

While this unfixed graphic nature of the company label appears not to have been a conscious Roots marketing deci-sion, it does reflect a refinement of the company's approach to lifestyle leisurewear. For example, while Nike – in many ways, the corporate Hyde to Roots' Jekyll – has reduced its logo to the single, inescapable and odious "Swoosh," Roots has been malleable, adaptive and – strange as it sounds in marketing terms – organic. Unlike the corporate conformity implied by the unchanging scimitar of the Swoosh, the any-way-is-okay rendering of the Roots name suggests a compa-ny that is loose, open and always ready to change. In other words, a nice place to belong to.

Most Roots crests are designed and produced by Roger Scannura and his team. Trained as a graphic designer, he spent three years in Europe learning embroidery design. Born in Malta, Scannura is also a flamenco guitar recording artist. Go figure.

Girls just wanna have abs:

sportswear pumps

up the volume.

Pump Up the Volume

Roots Sweats the Eighties

"You look mahvellous!"

— *Billy Crystal,* Saturday Night Live

In a strange but accurate case of dramatic symmetry, the Roots story can be neatly divided into three items of apparel, each introduced during a different decade, each representative of the era that produced it, and each one a reputation-making success. From the negative heel in the Seventies to the poorboy in the Nineties, Roots has taken a journey in fashion that literally ascends from toe to head. Act two of this drama of upward mobility features something in between. As far as Roots is concerned, the Eighties were really about one thing: sweatshirts. Sweatshirts with beavers on them. During that decade, Roots sold a gazillion of the things.

Before we examine how that happened, it might be profitable to wonder why it happened, to ask what it was about the era that permitted the Roots sweat to go boom. After all, sweatshirts were hardly new when the wildfire caught: they'd been around for at least a century, in more or

Jet-set sweat:

the mighty RBA takes off.

less the same style, and serving more or less the same function: to soak up the perspiration of people engaged in perspiration-producing activities. But in the Eighties, some set of sociological circumstances aligned themselves in such a way that a garment that was once considered a small step up from utilitarian underwear suddenly became — of all bleedin' things — a status symbol.

Naturally, it could only have happened during the Eighties. It was the age of abs, the pectoral period, the endorphin era. The what-me-worry Seventies time-shifted into the no-nonsense Eighties and suddenly everybody was taking orders from, of all role models, Olivia Newton-John. Indeed, where much of the pop cultural energy of the previous two decades implored us — through psychedelics, music, meditation or communal living — to get metaphysical, the *Grease* girl signalled the downshift into earthly practicality by slinking into spandex, surrounding herself with buff slabs of beefcake and singing "Let's get physical, physical..."

If you were still suffering temporal decompression (or, as likely, detoxification) from the Gimme Decade, the effects of the change could be downright disorienting. Suddenly, or so it seemed to those slow on the uptake, everybody was pumping up the volume and declaring war on cholesterol and body fat. Everywhere you looked on the cultural landscape, you were confronted by pectoral perfection: it was the time of Richard Simmons and Jane Fonda — the former Sixties radical and ultimate proof the revolution was as dead as Elvis — of aerobics and dancercize, of the rise of Rambos and

At Roots it's called a board meeting:
Green and Budman run for it
in Central Park, 1977.

Terminators, *Footloose*, *Fame* and *Flashdance*, Nautilus and Thighmaster, carbs and cardio.

Saturday Night Live's Hanz and Franz were promising to "pawmp" us up and, a few years earlier, the same show's Joe ("Are you from Jersey?") Piscopo traded in a credible comedy career to become a walking endorsement for bench-pressing. Speaking of Jersey, Bruce Springsteen, a skinny kid with a guitar and sneakers in the Seventies, became a virtual rock & roll *ubermensch* in the Eighties, with biceps as big as the state turnpike. It was a golden era for the physique and its conspicuous display, when the carefully sculpted human body became a commodity almost as lucrative as junk-bond trading. It was Madonna's moment, and a good time for Demi Moore to take it all off for the aptly named *Vanity Fair*.

It was a time when a whole new idiom infiltrated the popular lexicon, when the language of fitness beefed up conversation with hitherto arcane terminology: cardiovascular, aerobic, deltoids, endorphins and ectomorphic. If, as

President Reagan suggested at the decade's outset, it was indeed a "New Morning in America," then the day didn't get going until everyone had done a few push-ups.

Health, in other words, acquired a prominent toehold in the public consciousness, and so did its attendent manifestation, the cautionary recovery testimonial. In fact, almost as common were famous folk fessing up to past sins, indulgences and indiscretions, and all from an unambiguously repentant point of view. What amounted to a new secular, electronic revival meeting developed in conspicuous concurrence with the explosion in celebrity journalism: the public atonement, where stars appeal for forgiveness of past sins in the open confessional of the TV talk show. Indeed, just as Jane Fonda had traded politics for aerobics – and ex-radical Tom Hayden for media mogul Ted Turner – repentance of past bad behaviour became the ticket to televisual ubiquity for reformed reprobates. Dennis Hopper, once a poster boy for pharmaceutical intemperance, became the showbiz embodiment of Nancy Reagan's "Just Say No" campaign. Mick Jagger, the apotheosis of decadent rock & roll royalty in the Seventies, became the picture of pink – as they began to say – "wellness" in the Eighties; the better to prepare the world's most impervious rock & roll band for its reincarnation as the unstoppable touring machine, delivering oldies to the world's boomers.

The Stones' story is useful here, for not only would Keith Richards – the planet's most unlikely candidate for surviving the Seventies – make a prominent appearance on

'Pawmped' up: superstar of sweat Arnold Schwarzenegger conceals his pigeon chest behind Roots and Linda Sobel, Roots' director of special sales.

ROOTS!
FOR THE HOME TEAM

Roots

Roots' corporate client list, but his band's passage from bloodshot libertines to the step class of rock tours is a productive analogy of the shifting generational values that mark Roots' evolution from countercultural boutique to international leisurewear retailer.

If the general passion for conspicuous fitness in the Eighties can be seen as an indication that the youth-fixated and increasingly omnipotent boomers were staring mortality in the mirror for the very first time, Roots' success in that decade cannot be extrapolated solely from that fact. With their well-established policy of using their own interests and tastes as their only market research — yet another reflection of their generation's often breathtaking we-are-the-world confidence — Green and Budman were poised to crest the new wave like canoes on white water.

For with the shift from social to personal taken by boomer culture, the accessories of fitness acquired a previously unheard-of cultural and commercial status. Indeed, in one of the most significant fashion developments of the past half-century, athletic leisurewear became the hottest and hippest growth sector since teenagers discovered denim after World War II.

Keenly fitness-prone since childhood, Budman and Green were about to experience yet another profitable symbiosis of their own interests with that of the mass consumer. Long fascinated by the accoutrements of organized sport — team jackets and crests, sport jerseys and school sweaters — they were naturally willing to test the general application

No base left untouched.

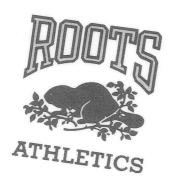

The RBA logo:
as ubiquitous in the
Eighties as bad hair. Almost.

of their personal interest in the public marketing laboratory of their store. In 1975, while shifting the Roots emphasis from shoes and leather accessories to a wider range of high-quality casual apparel, the partners stumbled across the Roots Beaver Athletics sweatshirt or, as it's called in Rootspeak, the mighty "R.B.A."

Had it happened to any company but Roots — an outfit that restores one's faith in timing, instinct and sheer luck — the story of the blockbusting R.B.A. would seem the product of pure public-relations whimsy.

It's 1975, two years after the Roots boys have opened their doors, and they are already millionaires because of the out-of-nowhere success of the negative heel. While the podiatric curiosity continues to sell like summer ice cream, the two know it's only a fad; and fads — like most sudden blossoms — come with a limited life. That's why, even a year or two before skepticism about the therapeutic properties of the negative heel would finish off the phenomenon, Michael Budman and Don Green were already looking for other stuff to sell.

As usual, they weren't looking in magazines, on runways or in market-research reports, they were looking at what happened to pass by the window of their immediate attention. On this day the two were working out — together as usual — on the leafy, tradition-steeped Toronto grounds of Upper Canada College (yet another emblem of high Canuckdom the ex-Motowners seem almost unnaturally drawn to) when they spotted the sweatshirt uniforms worn by the U.C.C. students engaged in phys-ed on the grounds.

Emblazoned with the collegiate insignia of this toniest of private educational institutions, and thus about as hip as three-piece pinstripes, the U.C.C. sweat tapped a well of positive associations for the Roots boys: it spoke of team effort and timelessness; health, fun and youth; and of the reassuring promise of something immutable. While it would be a full decade before these qualities would prove equally appealing to the general public, it was enough for their purposes — as it always has been — that *they* were inspired.

They moved immediately. After tracking down the Nova Scotia manufacturer of the U.C.C. sweats, they ordered a shipment of tops identical to the collegiate model in every respect save one: the crest on the left side of the chest bore the Roots corporate logo, a beaver's silhouette above the word "Roots." And the Roots sweat sold well enough to justify expanding the range of styles and colours.

By 1985, ten full years after Budman and Green saw the future on the U.C.C. grounds, the public appetite for athletic leisurewear was about to hit critical mass, and Roots decided to introduce the sweatshirt that would make the negative heel look like a slow starter: the now legendary Roots Beaver Athletics sweat, one of the most successful garments ever manufactured in the Dominion. Once again, success didn't come knocking, it broke down the door. "It started small," company executive Marshall Myles told *Toronto Life*. "A few sweatshirts in a basket in the corner,

Boys and the hood:

the Nineties.

113

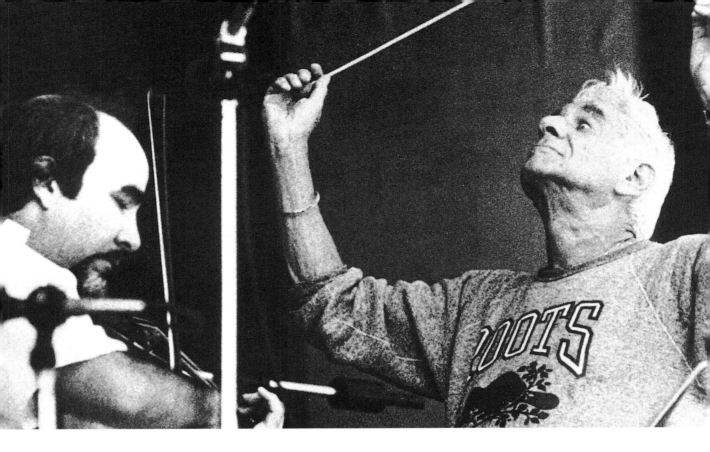

more for staff than anything else. And it evolved. The numbers became unbelievable. We couldn't give enough floor space, time, effort to it. I remember a Boxing Day at the Eaton Centre store, bringing product out from the back and being almost killed by customers."

Boasting a larger, front and centre logo (with "Roots" above the beaver and "Athletics" below), the sweatshirt became a casualwear monster. The company strained to keep up with the demand, and imitations were soon ubiquitous. By the early Nineties, the logo had appeared on more than a million garments.

There are many possible reasons for the mid-Eighties explosion of the R.B.A. By that time, the culture of fitness was in full, sweaty swing, probably reflecting the growing

health-consciousness of an aging baby boom. Then there is the mock-collegiate design of the logo itself. Budman and Green's passion for the symbols of institutional culture was hardly eccentric: this was an age when the need for some sense of communal belonging rose in reaction to the isolation of an increasingly technocratic and globalized culture. The R.B.A., with its graphic evocation of "team," made the wearer part of a virtual club. It was tradition you could buy, a sense of history and belonging instantly attainable for the price of a sweatshirt, smart-looking and well-made, reflecting the durability and timelessness evoked by the logo itself. It was teamwear in the absence of a real team, and thus a testament to the potency of the idea of team. The essence, that is, of something called team spirit.

above Sultans of sweat:
Mike and Don go to bat
for the mighty RBA.

opposite Leonard Bernstein
bears the beaver: from
symphonies of sweat . . .

. . . to pastures of perspiration:
Bobby Hull shows off his Roots.
His friend is unimpressed.

The royal coup: Prince Harry and his Dad in the shot seen round the world, 1998.

Yo! Highness!
Roots and Fame

"They're brash. When they made Roots shoes, they sent them to the Beatles' Paul McCartney. They were years ahead in seeing how important the media is to advertising."

— *Phil Nutt, Bata Industries, 1983*

"It's not one of **those** jackets is it? I can't **believe** this. This is **fantastic**! I don't even know you but I've just got to give you a hug!"

— *Michael J. Fox, on being presented a Roots Team Canada jacket by an interviewer, 1998*

"On November 5, 1989 Roots Pacific Centre Vancouver received Corizon Aquino as a customer. President Aquino was accompanied by her daughter, several private security guards and an entourage of political representatives. She stood outside our store and one of her aides briefed her about the impact Roots has had on Canadian culture. We greeted her upon entering and her daughter made a purchase of a Roots T-shirt. President Aquino was a most dignified customer and we were honoured by her visit. Unlike the Philippine's former First Lady, she didn't buy any shoes!"

— Roots Report!!, *in-store newsletter, 1989*

Unlikely, but it remains possible that, even as late as 1997, you might still have found someone, somewhere in Canada who wasn't aware of Roots. By 1998 — *early* 1998 — there was no such person on the planet. Now it's hard to imagine a completely Rootsless spot in the galaxy. Few would be surprised to see the Roots logo beamed back from the surface of Mars, or to see the new *2001* recast with a Roots beaver embossed on the monolith. For if any company had demonstrated an uncanny knack for being at the right place in the right time for max exposure, it was Roots.

Not that the company was exactly invisible before the Winter Games in Nagano, Japan. It's just that the eye-catching, funky-chic Canadian Olympic uniforms represented several years of savvy media strategizing by the company and the point at which public awareness simply went — speaking of outer space — nova. Oscar-nominated film director Atom Egoyan wore the cap on a section-front profile in *The Toronto Star*. Quincy Jones was presented with a Roots jacket on the Canadian music industry's Juno awards. Prime Minister Jean Chrétien was caught wearing his Roots in public, as was talk show host Rosie O'Donnell, and even the not-famous-for-being-hip Premier of Ontario, Mike Harris.

clockwise from top right
Roots wraps John Candy, Doug Gilmour, Boyz 2 Men, and Prime Minister Jean Chrétien – with Karl Kowalewski. Apparently, you're never too famous to appreciate free stuff.

Crowning achievement:
Prince William trades beats
with father, Charles.

When the Prince of Wales showed up with sons Harry and William at a press conference in Vancouver in March, not only was it one of the most widely covered photo-ops of the year, it was a Roots-op too: presented with Olympic-issue caps and jackets by the tireless Budman and Green, the Princes donned them for photos that were reproduced in newspapers around the world. In 1997, when the seventeen leaders of the Asia Pacific Economic Cooperation (APEC) association met in Vancouver, each was offered, and subsequently photographed in, a Roots jacket designed specifically for the occasion — President Bill Clinton's was even seen as the Chief of State boarded Air Force One for departure. "Spice World," my butt — by spring '98, the world seemed to belong to Roots.

Or at least to Canada, of which Roots is an increasingly prominent retail representative. Indeed, in the coming months, Canadians were as inescapable as Bill Clinton's sex scandals. In front of a billion viewers in March, Canadian-born director James Cameron beat fellow Canuck Egoyan for the Best Director Oscar — declaring himself "king of the world" — shortly after Quebec's Céline Dion sang her Oscar-winning song from the record-setting blockbuster *Titanic*. It was the night that the script for *Good Will Hunting*, written by Roots' models Ben Affleck and Matt Damon, snagged an Oscar of its own. In April, it was announced that Canadian journalist Kevin Newman had landed nothing less than the coveted hosting gig on *Good Morning America*, and Canadian singer Sarah McLachlan found herself on the cover of *Rolling Stone*. *The Truman Show*, a big-budget vehicle for the singular talents of former Torontonian Jim Carrey, opened in June. Indeed, at certain moments in the first half of 1998, so prominent were showbiz Canadians that it was possible to think that maybe the People of the Beaver were taking over.

Then again, if there were any feasible route to Canadian global domination, you couldn't find a more appropriate one than showbiz, a fact that Michael Budman and Don Green cottoned on to long before anyone else did.

Unfortunately for people who ponder such things, the issue of whether it was pluck or timing that led Roots to become one of the most celebrity-smart clothing manufacturers of the era may remain a conundrum. While it's clear

Briefly inescapable: The poorboy on Sarah McLachlan.

top left **Another funny Canuck: Martin Short. 1984.**

top right **We knew him when: Dan Aykroyd with Rosie Shuster in the Roots store in 1973.**

above **Aykroyd shirt-shopping in 1993.**

below **And celebrating the 25th in New York with Don and Mike.**

that the company has no shortage of chutzpah when it comes to slapping their products on the famous, the fact that the timing was so good renders all speculation moot.

One buddy present during those very first weeks of Roots business was a gifted young comedic sketch actor, formerly of Ottawa, named Dan Aykroyd, a part-time laundry-truck driver and performer at the newly opened Toronto Second City theatre. In fact, when an emergency trip to the airport for freshly shipped Roots product became necessary, it was Aykroyd's laundry truck that picked it all up.

Aykroyd, who within two years would become a bona fide boomer superstar on *Saturday Night Live*, wasn't the only future Not Ready for Prime Time player on the prehistoric Roots premises. Also helping around the shop was Gilda Radner, like Aykroyd a Second City performer by night and, like Green and Budman, an American expatriate and former Tamakwan.

Needless to say, with pals like this, Roots found itself in

prime strategic proximity to other future stars of comedy —
Martin Short, Eugene Levy, Catherine O'Hara, John Candy
and Lorne Michaels. Yet, while there's no doubt this played
a crucial role in the company's approach to celebrity promotion, it was really just adding steroids to muscle tissue waiting to be developed.

Like most of their generation — the first raised on
TV and rock & roll — Green and Budman grew up with a
healthy fascination and respect for pop culture's overachiev-
ers, especially those plucked from their own generational
pool. It did nothing to dampen matters that they were raised
in Detroit, home to Motown, Mitch Ryder and — thanks to
the red-and-white colours of the Red Wings and the cross-
border broadcasting of *Hockey Night in Canada* — a swell
place to fall in love with hockey. For these guys the only
stardom more luminous than music, movies or TV was sport.
Come to think of it, the peculiar culture of Roots might be
the intersection of Smokey Robinson and Gordie Howe.

Gary Leeman, Wendell Clark, Russ
Courtnall and fans, including
Michael Budman, 1986.

Hometown soul: Roots mourns
Motown's Marvin Gaye.

The other business of fame:
exclusive label clothing
for showbiz.

There was more. Green's cousin was (and remains) future blockbuster producer Jerry Bruckheimer (*Top Gun, Con Air, Armageddon*), which might suggest that a certain showbiz brashness runs in the family.

Still, for all that these connections and inclinations contributed to the celebrity cachet of Roots, the right-place-right-time factor cannot be underestimated, especially in the Toronto-in-1973 sense. Still a pretty quiet and WASPY town when the two hirsute kids from Detroit arrived, Toronto was on the cusp of transformation from provincial Upper Canadian capital (where it was illegal to open stores or drink in public on Sunday) to the bustling multicultural metropolis it is today. The once-vibrant music scene in Yorkville had been largely silenced due to civic paranoia about the hippie apocalypse. There was no CN Tower to signify the city's heaven-storming ambitions, and the only sports to speak of were the local CFL team and the beloved Maple Leafs. The SkyDome didn't exist, and the Victorian splendour of the Royal York Hotel played prominently in the city's otherwise modest skyline. Montreal had the monopoly on Canadian urban cool and, if Toronto were famous for anything internationally, it was as a haven for draft dodgers and home of Wayne and Shuster.

Within five years, all that would change. Just as the skyline rose in an architectural chorus to the space-age phallus of the CN Tower, so the city's staid and insular self-image was boosted by a number of unabashedly immodest developments: the launch of Moses Znaimer's internationally

First Roots sponsored a jock-doc called *Les Canadiens*. Then they threw a party, and *everybody* came.

top Budman, Green, Ronald Corey, Aurel Joliet, Guy Lafleur, Jean Belliveau, and Maurice Richard. *middle* Tony Bennett and Michael. *bottom* Mordecai Richler and son, Jake, with Michael and Diane.

imitated City-TV experiment in 1972; the boom in TV and
movie production occasioned by a 100 percent federal tax
deduction for Canadian productions; the launching of an
international film festival in 1975; and the incredible success of Lorne Michaels' Canuck-filled *Saturday Night Live*,
which some saw as nothing less than the successful storming
of the Big Apple itself. Collectively, these events not only
granted Toronto a long-overdue dose of cool cosmopoli-
tanism, they came to signify a phrase as unavoidable as the
Tower itself: "world class."

For the rise of Roots, and particularly the company's
developing approach to celebrity exploitation, these contex-
tual conditions cannot be underestimated. Aykroyd and
Radner's sudden fame gave Budman and Green instant
entrée to what would soon be the hippest celebrity scene
on the continent: the New York *Saturday Night Live* party
circuit. The two entrepreneurs were virtual fixtures at the
infamous after-hours club run by Aykroyd, as were their
large bags of Roots freebies. (As friend Marcus O'Hara —
scenemaker, entrepreneur and brother of SCTV's Catherine —
once told an interviewer, the two were known in New York
as "the Santa Clauses of the North.")

Moreover, the explosion of Toronto film and TV production in Toronto meant the two were in valuable proximity to an industry that was a high-profile purchaser of
customized clothing, and would eventually prove to be one
of Roots' most lucrative professional patrons. The Festival
of Festivals, as the Toronto International Film Festival was

clockwise from top left John Belushi and Don; Toots Hibbert of the Maytals; Gary Shandling; Stevie Wonder with Don and Greenlet; Chris Farley; Mark Wahlberg; Marty Kove; Jeannie Beker poses questions at the *Indian Summer* movie party; and Robert Plant.

above **Phantom merchandise: more practical than a cape.**

opposite **By Special Appointment to Showbiz: Roots reaches for the stars.**

then called, offered more celebrity access: the event, which now ranks among the world's foremost festivals, brought some of the biggest stars and directors to the shores of chilly Lake Ontario. High-profile sponsors of the event virtually from the beginning, Budman and Green took stars for exclusive, not to mention free, after-hours Roots shopping sprees while in town.

The arrival of the Blue Jays baseball franchise in 1977 marks another crucial development in Toronto's public life that Roots capitalized on. The Jays gave the city both a new team to, um, root for, and another boost to its steadily inflating civic self-image. Seemingly overnight, Toronto became a baseball town, and Roots rushed the playing field almost as quickly as the fans. Without a doubt, the company's prominence as outfitters to the world of sport was developed simultaneously with Toronto's own quickly swelling jockdom: by the mid-Eighties, the city had built a state-of-the-art sports dome, and by the mid-Nineties, it even had its own NBA basketball franchise. A company run by a pair of starstruck sports fans might eventually have developed the professional sport connections it did without the Toronto jock-boom of the past twenty-five years, but there's no doubt that being there at the right time certainly didn't hurt.

And yet, while Roots' celebrity strategy goes back virtually to day one, actual endorsement advertising did not appear until years later. A number of possibilities are suggested by this. First, that Budman and Green — who featured

Brushed Twill
Baseball Cap

Baseball Jacket

Dylan Car Coat

Twill/Nubuk
Baseball Cap

Shirt

O INTERNATIONAL
FESTIVAL

Jean Jacket
with Leather Collar

THE
Tonight
SHOW
WITH
JAY LENO

Baseball Shirt

SATURDAY
Night
LIVE

HBO
BOXING

Melton/Leather
Baseball Cap

SNL

Twill
l Cap

Twill Nubuk
Baseball Cap

Roots
MADE IN CANADA

A family business: down-to-earth stars
and Roots loved ones snuggle.

Sting sports Roots quality: the
jacket will outlast the hair.

themselves in early ads more often than their famous friends — were initially drawn to celebrities as fans, not equals. Second, that they recognized the long-term public relations importance of nurturing relationships with high-visibility athletes and showbiz types. Third, that they weren't thinking about endorsement in the conventional sense at the beginning. Instead, they saw two other potential Roots' boons from the care and feeding of celestial bodies: the promotional importance of having clothes appear *naturally* on famous people — which suggests the stars *like* to wear them and not that they're being paid to do it — and the enormous profit potential in producing custom casual wear for the entertainment industries.

There's little doubt that, as lifelong friends who are also successful business partners, they understood the benefits of developing high-profile relationships based on loyalty and respect. This is another manifestation of the company's family and friendship orientation, which dictates that strong

and happy alliances – family, friends, business partners and stars – are necessary to success and longevity in any venture. Possibly this explains why, when the company did call in some famous friends and acquaintances to be Roots models, there was often such a relaxed and homey feel about the result.

The photos of Jason Priestley, Robbie Robertson, Elvis Stojko and Kelly Lynch conveyed nothing so much as a casual and unrehearsed comfort in their Roots, an impression bolstered by the inclusion of stars' kids and pets. (Compare this to, say, Michael Jordan's endorsement arrangement with Nike, which seems about as warm, fuzzy and genuine as a hostile corporate takeover.) Moreover, the Roots celebrities are also seen to endorse the highly publicized *values* of the company – literacy, environmentalism, fitness – as well as its lifestyle philosophies. Comfort, relaxation, family and commitment to good causes – the human side of stardom. It not only wears, it cares.

Casual, or its appearance, is the key here, which may explain why Roots has simply given clothes to famous folk – to have them appear in the stuff in public by choice: Rosie O'Donnell wore her Roots jacket because she *wanted* to. *Choosing* someone famous to wear your line of clothing is one way to use fame as a promotional tool, but for Roots, it's not nearly as effective as having someone famous *choose you*. In the casual-chic world of Roots, that's the ultimate endorsement. And it's worked: ever since Paul McCartney was caught flipping a Frisbee while wearing negative heels

top to bottom **Jason Priestley and Swifty Lugnuts; Kelly Lynch and daughter, Shane; Robbie Robertson and son, Sebastian.**

Heat sensing:
top **Atom Egoyan, before his Oscar nomination;**
below **Spike Lee, Billy Baldwin and David Foster.**
opposite **While multiple Oscar-winner *Good Will Hunting* was being shot in Toronto, Roots had the foresight to dress and shoot its imminently hot cast. Matt Damon, Ben and Casey Affleck.**

in a 1976 issue of *People*, Roots clothes have appeared *by choice* on celebs from David Bowie and Elton John to Janet Jackson and Rosie O'Donnell.

Nurturing relationships with the entertainment industry has also yielded Roots one of its most abiding, envied and lucrative non-commercial sidelines – supplying exclusive label clothing for movies (*Malcolm X*, *Forrest Gump*, *Pulp Fiction*), TV shows (*Saturday Night Live*, *Seinfeld*, *Oprah Winfrey*), big-ticket theatrical extravaganzas (*The Phantom of the Opera*, *Miss Saigon*, *Second City*) and rock tours (Janet Jackson, Keith Richards, Bryan Adams).

Any way you parse it, it's hard to disentangle the fame and success of Roots from the fame and success of the friends it has cultivated. Indeed, one of the company's most enviable talents has been the ability to detect fame just before it blasts off, as it clearly did when it featured Affleck, Damon and Minnie Driver, who were in Toronto filming *Good Will Hunting*, before most people knew who they were.

Budman told an interviewer on the company's tenth anniversary in 1983 that Roots just got lucky: "We knew a lot of people before they were famous."

Lucky, maybe, and more than a little smart.

132

The Eaton Centre, Toronto, 1994.

Cottage Industry
Roots Does Business

"Don Green and Michael Budman are the Little Jack
Horners of the fashion industry. They have their
thumbs in every pie and always seem to score a plum."
— *Syndicated newspaper profile of Roots, 1998*

"Public taste is evolving in our direction."
— *Michael Budman, 1995*

Of all the commodities available at the Eaton Centre in
downtown Toronto, tranquillity is not among them. One of
the first large inner-city malls to open in Canada — and the
prototype for similar retail environments across the country
— the multi-tiered, two-block-long shopping emporium is
nothing short of an audio-visual riot. Below the impervious
Canada geese suspended by artist Michael Snow from the
glass and steel-domed ceiling, three strata of tightly packed
stores compete for the attention of overwhelmed shoppers
as escalators deliver them from one level of non-essential
temptation to another.

The Backstreet Boys at the Roots lodge, er, store.

Opened to the public just four years after Roots opened its doors a couple of kilometres due north, the Eaton Centre is — along with the CN Tower and the bank towers — an architectural monument to the skyscraping heyday of Toronto's World Class period. It is oversized, immodest, unsubtle and in a continuous state of anxious hustle. Indeed, when busy (as it often is), it feels like nothing so much as the environmental equivalent of a constantly zapping TV. Not, that is, a place of peace, reflection or back-to-basics simplicity. Until you find yourself at the Roots store on the third level — far up from the retail rabble around the food court, with the posh stuff. Located on a *faux* boulevard replete with benches and trees, and flanked by such other retail emporia as Banana Republic and Nine West, the Eaton Centre Roots store was, before its renovation in 1998, an eye-catching study in contrast. In an environment that stresses downtown slick and shiny — chrome, steel, glass and mirrors — the Roots store stuck out like a canoe on a superhighway. Large wooden beams framed the storefront, like the entrance to some remote lakeside lodge, and the inside was similarly woodsy: the high-quality, natural-fabric Rootswear, with its comfortingly restricted palette of warm colours, was smartly displayed on polished pine surfaces, under a ceiling of the kind of beams one usually associates with summer-camp mess halls and cottages built by caring and calloused hands. The only joint in the mall that went with the geese.

This impression of homespun, rustic welcome was augmented by the fleecy soft-sell of those famously happy Roots poster faces: actor and Academy-award winning screenwriter Matt Damon, figure skater Elvis Stojko, gold-medalist snowboarder Ross Rebagliati — all looking genuinely happy to be wrapped in Roots, as if it were the most, um, natural thing in the world.

Which is the point. An oasis of 100 percent cotton, cottage country serenity in a desert of urban hard-sell, the Eaton Centre Roots outlet — like Roots outlets in similar retail environments everywhere — is the physical embodiment of the Roots retail philosophy: stand apart from competition with the low-key sincerity of your pitch, and offer the product as part of a lifestyle. Despite — or indeed *because* —

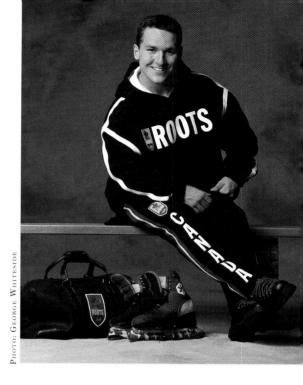

PHOTO: GEORGE WHITESIDE

above **Elvis in the building.**

below **Former brat-packer Molly Ringwald grows into Roots.**

PHOTO: GREG GORMAN

of the surfeit of consumer choice available in places like the Eaton Centre, Roots stores aim to create the impression of being the only place you want to be, a shelter from the shopping storm.

For as important as the unimpeachable craft and quality of the individual products are, is the lifestyle they supplement that's what's really being sold. And it's in this that Roots really stands apart from competitors. (Asked once whether companies such as Banana Republic and The Gap, which also sell limited lines of high-quality leisurewear, pose any threat to Roots' marketplace distinctiveness, Michael Budman shrugged: "There's nobody that directly competes with us.")

While it's pretty hard to convince anybody — let alone anybody caught in the hustle of the Eaton Centre — that what they *really* need is a new leather knapsack or Roots watch, you can always sell people a better way of living, to which — who knows? — a leather knapsack or Roots watch might be an entrée.

This is what Roots has really been marketing since those early, alternative-lifestyle ads for the negative-heel shoe: a better world — cleaner, healthier and more conscientious — through "proper" shopping. Needless to say, it's a dicey proposition. Few would deny that consumer culture is a key contributor to the social and environmental problems Roots stands firmly against — but only if you make the mistake of confusing marketing with real life. As the real product of any successful retailing is fantasy — the creation of an apparent need where an actual one doesn't exist — Roots lifestyle marketing is inspirational. Since the goods themselves are offered as part of a program for better living — a fact stressed in stores by posters, photos and newsletters printed on recycled paper — Roots lets you have the best of both worlds. You can buy something sharp that you probably don't really need (as if "need" ever had anything to do with effective marketing), *and* you can feel you've contributed to something positive. It's altruism through self-gratification, a mighty alluring pitch.

Not that Roots — always more the refiner than innovator — is not alone in such indulging. At one end of the responsible-consumption scale is The Body Shop, the British cosmetics giant that shares at least three of Roots' marketing philosophies. It positions itself as a kinder, gentler and more thoughtful retailing alternative. It aggressively promotes its products (which, like Roots', emphasize quality over economy) as accessories to a healthier and more environmentally responsible lifestyle, and it incorporates the support of certain causes into its marketing. "Against Animal Testing" appears on just about every fully recyclable cosmetic receptacle The Body Shop sells, and a 1998 magazine ad campaign implored consumers to buy Body Shop-issue Amnesty International candles to support the release of international prisoners of conscience. "While there's no guarantee that prisoners of conscience will be freed," the ad cautiously states, "doing nothing guarantees they won't." Here conscience and consumption are synonymous: *not* buying

Wear that cares: Roots goes green.

below Well, maybe worry a little:
Bobby McFerrin in Roots ECO-T.

A Roots t-shirt
is the
best built t-shirt
in the world.

A Roots t-shirt

makes a statement.

This spring
get the message
from Roots.

Roots
CARES

contributes to the problem. This particular sell is slightly harder than one would expect from Roots, which has tended to shy away from overtly political campaigns, but the underlying principle remains the same: one can save the world, or at least make it a little nicer, and still shop.

This is not to suggest that Roots' commitment to any causes it supports — literacy, environmentalism, fitness — is less than genuine, just that the commitment itself is an attractive and essential element of what Roots sells: not just clothes, but the hint of a better world to wear them in.

Indeed, it's something that has been part of the Roots pitch since the beginning. "You'll find genuine Roots only at Roots stores throughout Canada, the U.S. and Europe," a 1975 negative-heel ad tells you. "And you'll find these stores as comfortable an experience as Roots themselves. They don't look like ordinary shoe stores. And the sales people don't act like ordinary sales people. You'll find them wearing what they're selling and servicing what they sell."

Packed into those few lines of non-hyperbolic, gently lobbed ad copy is the business philosophy of Roots. There is the soft-sell sincerity, the take-it-or-leave-it tone that conveys complete confidence in the quality of the product. (The ad's first line reads: "Everyone who owns a pair of Roots can skip this.") There is the granola-holistic unity of product, salesperson and store environment, a unity that implicitly suggests this isn't just about buying shoes, it's about choosing a way of life and, just as important, the *freedom* to make that choice. "They don't look like ordinary shoes," the ad modestly

Pause for a cause
that anyone would buy into.

Family stitching: the Kowalewskis in the plant; (l-r) Henry, Jan, Richard and Karl. 1976.

admits, "which bothers some people. But, if you're secure enough to deal with a few characters who want to know why you're wearing those 'funny-looking shoes', you're going to love the comfort of Roots." And if you're not that secure, it's *your* problem.

The emphasis on choice — lifestyle over product, comfort over fashion — has remained crucial to Roots' consumer appeal. In the same ad is a retro-rustic photographic portrait of the Kowalewski family, which — along with the information that the staff wears what they sell — stresses that Roots isn't just an ordinary store, it's a store run by people who believe in their product, who are living the lifestyle the negative heel signifies. "John Kowalewski and his sons are custom bootmakers," we're told. "They built the first pair of Roots with the same care they poured into the expensive shoes they had made for exclusive boutiques."

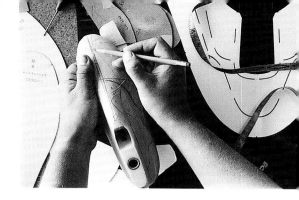

Just as the ad is confident that you the customer will do, the Kowalewskis *chose* Roots. The "exclusive boutique" days are over.

"Today they run the Roots factory in Toronto where every pair of Roots is made. They pick the leather, set the standards, supervise the whole operation.

"They are of the old school."

Family, tradition, quality and care — all packed into the lowly negative-heel shoe, and the qualities Roots has promoted for twenty-five years.

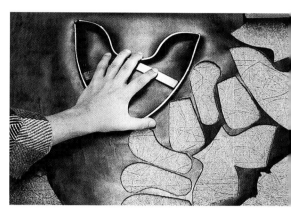

A big hand for the foot: crafting Roots shoes.

In a world where everyone seems to be selling *something* — from toothpaste to political platforms — it's easy to dismiss Roots' claims to customer-first commitment as just so much hooey. After all, children are used to promote big banks, Dylan sells computers and RRSPS, nationalism is tapped to sell beer, and even diet soft drinks somehow denote "lifestyle." Indeed, even the fact that Roots was pushing the alternative-lifestyle buttons before most others doesn't preclude one's reasonable suspicions of calculation. It only means they thought of it before the other guys did.

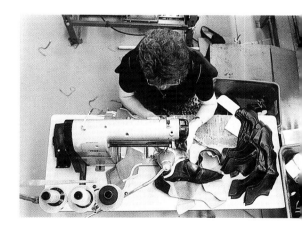

But maybe the issue isn't calculation — for what successful marketing enterprise isn't guilty of that? — but whether the claims made are legitimate or not. In other words, when Roots talks things like quality, tradition, comfort and commitment to a happier, healthier world, does it walk what it talks?

To answer this, one needs to return to the company's origins, for it is there one sees the seeds of the marketing

Hands-on dad: Al Budman offers fatherly advice to a fledgling enterprise.

(2)

Now: How do you cure this serious situation? Very Simple!!

A. Concentrate more than ever with extra effort to pay your suppliers. ... the suppliers ...andise you ... shelves + ... you have ... 75% of ... go to the ... a must. ... plant ... what ... & apply ... Rating ... is one ... special ... that ... 25% off

11/30/84 60° Cool

Dear Michael:

I am writing a synopsis of the meeting we had at the Beverly Hills Hotel.

Topic #A Accounts Payable: 30-60-90-120

Important!! If your accounts payable ... 60 Days then ... trouble. ... start to Concentrate ... 30 Days. Thats ... you reach 90 Days ... real trouble. ... technically you ... a Cash basis: ... spread; 90 days ... able & Cash Basis of Roots is too far apart for Comfort.

Too Much Pressure:

... items ... on a ... & you ... accts ... starts ... Saturday) 198... Conflict ... Sales + apply the ... accts Payable

Dec 27 Wed
" 28 Thur
" 29 Fri
" 30 Sat
" 31 Monday
1984

Jan 1 Tues
" 2 Wed
" 3 Thur
" 4 Fri
" 5 Sat
1985

Eleven Day Sale =

The supplier is more important than the Banker —

There is no reason why you should have acct's Payable pressure when it's not necessary

106-43
106-44
106-45

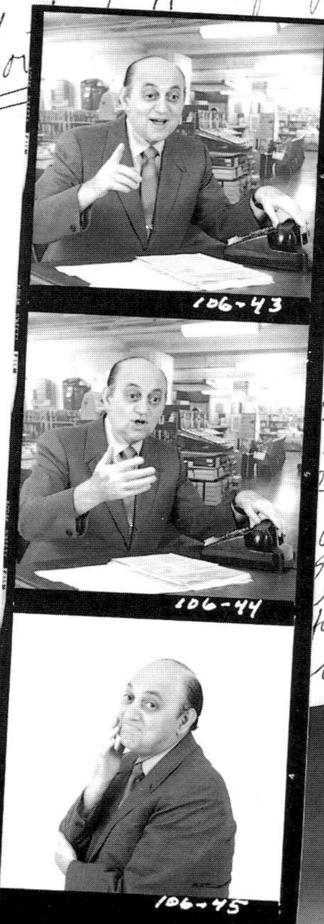

philosophy sown. One sees, for example, the exceptional longevity of Budman and Green's friend cum partnership, and the familial loyalty that marks both their childhoods. (One also sees that both come from a tradition of successful salesmanship: Budman's father was a successful aluminum-siding salesman, Green's father an eminent Motown auto-parts manufacturer. Moreover, both fathers played key roles in developing and advising the company in its early years: it's as Roots as the beaver logo.) You see the commitment to certain causes from the outset, and the cautiously hands-on approach of the owners: nothing is introduced that would not be worn in public by at least one member of the Green or Budman family.

Don, takin' care of business.

Indeed, the entire corporate structure and practice of Roots, a completely vertically integrated private company, is based on the principle that size matters: in this case the smaller the more manageable, and the more manageable the better. (The company has attracted more than its fair share of acquisitory interest — most famously on the part of the clothing giant Dylex — but as Green said of the Dylex offer, "They wanted to buy half the business. It didn't feel right. We didn't want that as our reality. We love the independence of this thing.")

What thus impresses the outside observer about the company, which has $150 million in annual sales and now competes in several markets worldwide, is the relative simplicity of the operation, and how that modesty in business practice reflects what Roots promotes in lifestyle.

Meet the boss: Budman, Green
and the factory.

The dawn of deep discounting:
Denyse Tremblay and Don Green
at their first factory sale,
in the early Nineties.

For example, refusing all offers that might compromise Budman and Green's ownership and control is a strategy to maintain managerial autonomy, but it also ensures that the company's public commitment to quality and "Made in Canada" production are not compromised by outside interests. (Ninety-five percent of Roots products are made in Canada under Green and Budman's personal supervision, with the production of leatherware still overseen by the Kowalewskis.) Moreover, Roots non-unionized employees work according to the "rink system," practised most famously by Toyota, which ostensibly promotes quality and short-circuits boredom by rotating people's roles in the manufacturing process. (Presumably, it also short-circuits the dissatisfaction likely to lead to union organization.) As a production method, this may be more costly and time-consuming than the Detroit-born assembly line model; but it helps maintain employee contentment and loyalty while ensuring that claims of craftsmanship and environmental responsibility are not mere claims.

When one visits the factory, the first thing one experiences — after having your shoes and knapsack quality-checked by Budman — is a personal, Kowalewski-guided tour of the facility. Everything is open to scrutiny and observation. Unlike Nike, for example, which manufactures athleticwear in the cheap-labour Third World, Roots would seem to have no reason to dread a surprise visit from corporate gadfly Michael (*Downsize This*) Moore. Moreover, "Made in Canada" is more than just smart P.R. That this stuff *is* made in Canada clearly appeals to the modest but proud sensibility of the country itself.

God knows, if you insist on being as successful in Canada as Roots is, it's prudent to be as modest about success as you are frank about failure, and Roots, whose public persona is still largely synonymous with its owners, is both. In retail terms, Roots transmits Canuckthink, which says "We might not be the biggest, but we're the best we can be under the circumstances." Or, "We may not be perfect, but godammit we're trying." As a method of doing business anywhere but in Canada — and particularly the U.S. — it would seem an absurdly naive approach to retail empire-building. But in Canada, it's nothing short of a stroke of marketing genius. Even better, these guys seem to *mean* it.

Indeed, one of the secrets to longevity in the Canadian imagination is not to seem overly successful or self-inflated, for immodesty is taken for a sure sign that you're probably not a true Canadian. Nothing if not sensitive to the fact that their birth certificates were printed *down there*, Roots'

**The doomed Aspen Lodge:
hopes snowed under.**

co-owners may play the maple leaf card a little too often and aggressively, but such an accusation reveals as much about the peculiar, unforgiving Canadian pathology as it does about anything concerning Roots. Damned if you don't wave the maple, darned if you do.

While Green and Budman have made a management principle out of accentuating the positive, they're Canadian enough to fess up to failures too: "You'll hear no excuses from the locker room" is typical Budmanese. Not only does self-deprecation sit well with their adopted culture, it helps alleviate the somewhat irritating assumption that nothing goes wrong for Roots, that the company is blessed with a bottomless supply of groovy karma. (It isn't: it's just blessed with *more* groovy karma than most of us are entitled to.)

Still, while stretches of treacherous whitewater have appeared, they haven't tipped the canoe. There was the lawsuit, for example, that saw the company investing a fortune to protect itself from charges of copyright infringement: turns out there was a New Jersey clothing company called Roots, and the eventual settlement (Green and Budman won) took ten years, three court hearings and a million dollars (U.S.).

Believe it or not, there have been dubious, ill-advised and costly miscalculations: a line of surfwear that tanked; and the Roots ski lodge and resort planned for Aspen, Colorado, a grandiose pipedream of a total Roots lifestyle environment – think "Beaverworld" – that eventually went downhill faster and harder than Wile E. Coyote on skis. There was the well-regarded but costly and short-lived *Passion* venture, a Roots-funded foray into English-language European magazine publishing that coincided with the company's entrée into the intensely competitive – and insular – Parisian fashion market. A success in terms of reputation-building, the mag nevertheless folded.

There was the recession of the early Nineties, a global economic downturn that, combined with the GST and a surge in cross-border shopping, hit the privately owned company hard.

above **Passion over reason: Paris remains a nice place to visit.**

below **Troubled waters.**

White water for Roots

Michael and Don with Irwin
Green, who helped navigate
the company through the
"white water" of the recession.

But then, in another display of karmic grooviness, the downturn was reversed: except for 1991, the company has seen an increase in annual profits every year.

As for the future, it's no easier today to predict where Roots might go, or how successfully, than it would have been on that day in 1973 when the first seven pairs of negative heels walked out the door. After all, this is a company run by two guys who share a mystical faith in whay they call "the gut", and who can say how long that will continue to growl profitably? All that would seem certain is that any predictions would probably fail. The negative heel defied just about any kind of logic you'd care to apply, and the success of the R.B.A. sweatshirt took everyone — including its manufacturers — by complete surprise. Ditto the poorboy, about which even Budman was harbouring serious doubts just a week before it marched proudly into Nagano. Yet the fashion craze many said would burn off with the arrival of spring was seen as late as July in the *New York Times Magazine*, adorning the precociously successful head of ultra-hip *South Park* creator Trey Parker. Then consider Manhattan's SoHo, site of what many called the riskiest Roots franchise opening ever, the heart of the most unforgiving fashion centre this side of the Left Bank. Roots opened there in June 1998, and by late summer the store was thriving.

So go figure. Only this much is certain: provided both corporate control and "gut" rumbling remain firmly under Michael Budman and Don Green's control — a growing challenge as the empire expands — the canoe looks set to remain upright for some time to come.

150

PHOTO FACING PAGE: NIGEL DIXON

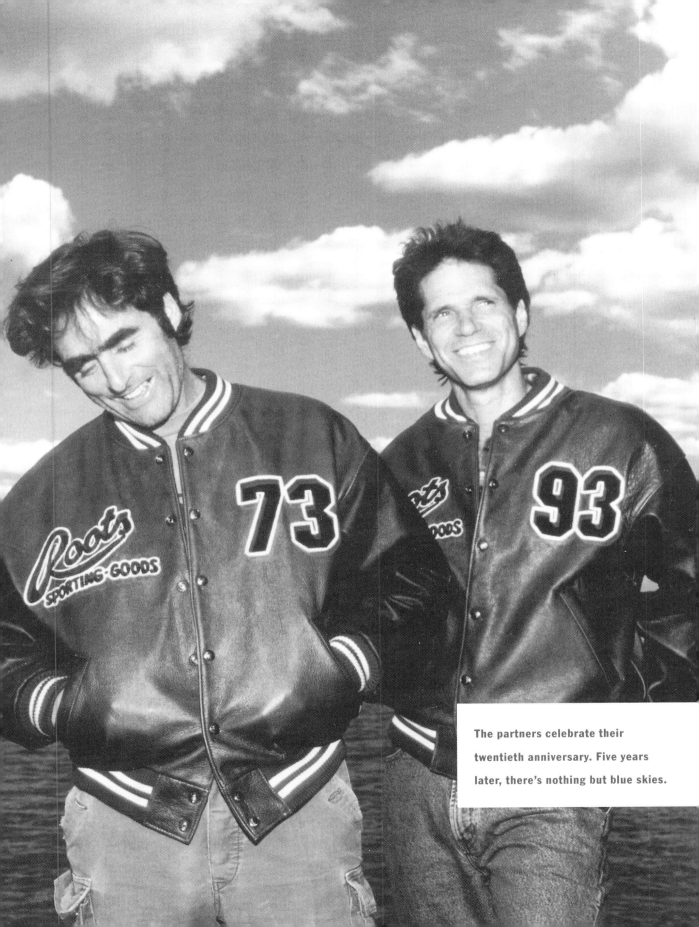

The partners celebrate their twentieth anniversary. Five years later, there's nothing but blue skies.

Photo Credits

8: Canadian Olympic Association/F. Scott Grant. 10: (top, bottom) CP Photo/Fred Chartrand. 11: Canadian Press CP/Ryan Remiorz. 13: (top, bottom) Canadian Olympic Association. 17: Canadian Olympic Association/Mike Ridewood. 19, 27: Gaynor Fitzpatrick. 20: Toronto Sun/Greig Reekie. 46: Steve Wahl. 48: CP Laserphoto/Paul Chiasson. 53, 54, 60 (middle), 61: Camp Tamakwa Archive. 64, 70: Photofest. 68, 72, 75, 78, 108, 134: Toronto Sun. 71, 76 (top): Covers courtesy of Linda Gustafson and David Hamilton. 74: Dick Hemingway. 77: CP Photo. 79, 94, 95 (top, bottom): Dick Hemingway. 82: Reuters/John Hryniuk/Archive Photos. 91, 151: Nigel Dixon. 92: Associated Press AP. 96: Toronto Star/R. Bull.
97: Toronto Sun/Mark O'Neill. 104: Canadian Press CP/John Lehmann. 107: Lyn Goldsmith. 116: Canadian Press CP/Chuck Stoody. 120: Reuters/Lyle Stafford/Archive Photos. 131 (top, middle, bottom), 133, 137 (bottom): Greg Gorman. 137 (top): George Whiteside.

All other photos are from the Roots Archive.